Characterization and Applications of Metal Ferrite Nanocomposites

Characterization and Applications of Metal Ferrite Nanocomposites

Editor

Thomas Dippong

MDPI • Basel • Beijing • Wuhan • Barcelona • Belgrade • Manchester • Tokyo • Cluj • Tianjin

Editor
Thomas Dippong
Technical University Cluj Napoca
Romania

Editorial Office
MDPI
St. Alban-Anlage 66
4052 Basel, Switzerland

This is a reprint of articles from the Special Issue published online in the open access journal *Nanomaterials* (ISSN 2079-4991) (available at: https://www.mdpi.com/journal/nanomaterials/special_issues/ferrite_nano).

For citation purposes, cite each article independently as indicated on the article page online and as indicated below:

LastName, A.A.; LastName, B.B.; LastName, C.C. Article Title. *Journal Name* **Year**, *Volume Number*, Page Range.

ISBN 978-3-0365-3074-1 (Hbk)
ISBN 978-3-0365-3075-8 (PDF)

Contents

About the Editor

Thomas Dippong (Associate Professor, Doctor at the Technical University of Cluj Napoca) is a chemical engineer with a PhD in chemistry. His current research activities are related to nanoparticles for various applications, as part of ongoing research in partnership with the Technical University of Cluj-Napoca within the field of nanomaterials. He is an expert in analytical chemistry, organic/inorganic chemistry, heat treatment, instrumental analysis, and synthesis of nanomaterials. He has published 118 papers (64 in WOS journals (20 Q1, 11 Q2 and 23 Q3), 54 in other national and international journals), 890 citations, h-index: 25 (WoS) and 34 communications at national and international conferences (ICTAC 14 Brazil, ESTAC Brasov, JTACC Budapest, CEEC-TAC5 Roma). He has also published one book at international publishing houses and 15 books at national publishing houses (five as unique and seven as first author). He has been the contract manager for projects totalling a budget of EUR 4 million and is currently an active member of four other projects.

Editorial

Characterization and Applications of Metal Ferrite Nanocomposites

Thomas Dippong

Faculty of Science, Technical University of Cluj-Napoca, 76 Victoriei Street, 430122 Baia Mare, Romania; thomas.dippong@cunbm.utcluj.ro

Citation: Dippong, T. Characterization and Applications of Metal Ferrite Nanocomposites. *Nanomaterials* **2022**, *12*, 107. https://doi.org/10.3390/nano12010107

Received: 18 December 2021
Accepted: 27 December 2021
Published: 30 December 2021

Publisher's Note: MDPI stays neutral with regard to jurisdictional claims in published maps and institutional affiliations.

Introduction

In recent years, nanosized spinel-type ferrites emerged as an important class of nanomaterials due to their high electrical resistivity, low eddy current loss, structural stability, large permeability at high frequency, high coercivity, high cubic magnetocrystalline anisotropy, good mechanical hardness, and chemical stability. Thus, research dedicated to the development and characterization of such nanomaterials, the development of cost-effective, eco-friendly synthesis methods, and finding new applications for existing materials has received considerable attention. Metal ferrites MFe_2O_4 (M = Mn^{2+}, Co^{2+}, Ni^{2+}, Mg^{2+}, Zn^{2+}) have been promoted as a novel group of versatile nanomaterials due to their tunable magnetic, electrical and optical properties together with good thermal and chemical stability. They are suitable for a wide range of potential applications in photoluminescence, catalysis, photocatalysis, humidity-sensors, gas sensors, biosensors, information storage, permanent magnets, transformer cores, radiofrequency circuits, waveguide isolators, hybrid supercapacitors, ferrofluids, inductors, converters, antennas, biocompatible magnetic-fluids, magnetic drug delivery, magnetic refrigeration, microwave absorbers, water decontamination and medical imaging, ceramics pigment, corrosion protection, antimicrobial agents, and biomedicine (hyperthermia) [1–7].

This Special Issue focuses on ferrite-based nanomaterial synthesis and characterization including (i) synthesis; (ii) advanced chemical and physical characterization of structure and properties; (iii) magnetic behavior; (iv) computational and theoretical studies of reaction mechanisms, kinetics, and thermodynamics; (v) applications of nanomaterials in environmental, biological, catalytic, medical, cultural heritage, food, geochemical, polymer, and materials science.

The control of the morphology and magnetic properties of ferrite nanoparticles (NPs) is critical for the synthesis of compatible materials for different applications [8,9]. In this Special Issue, Duong et al. [10] reported the synthesis of $CoFe_2O_4$ NPs by a solvothermal method using cobalt nitrate and iron nitrate as precursors, and oleic acid as a surfactant. Additionally, the effect of reaction time, reaction temperature, and oleic acid concentration on the properties of $CoFe_2O_4$ nanoparticles was investigated. The obtained results indicated that the oleic acid concentration plays an important role in controlling the morphology and properties of the $CoFe_2O_4$ NPs. The obtained high-quality $CoFe_2O_4$ NPs with improved magnetic performance are a potential candidate for many applications, such as bio-separation, magnetic resonance imaging, biosensors, drug delivery, and magnetic hyperthermia [10].

Due to its excellent stability, non-toxicity, biocompatibility, drug loading capacity, and water dispersibility, the mesoporous SiO_2 enhances the stability of NPs in water, improves chemical stability and minimizes the agglomeration of NPs, without influencing their magnetic and dielectric properties [1,8,9]. In this Special Issue, the effect of SiO_2 embedding on the production of single-phase ferrites, as well as on the structure, morphology and magnetic properties of $(Zn_{0.6}Mn_{0.4}Fe_2O_4)_\delta(SiO_2)_{100-\delta}$ (δ = 0–100%) NPs synthesized by the sol–gel method and annealed at different temperatures was reported by Dippong et al. [2].

1

The obtained results indicated that the preparation route strongly influences the particle sizes, and especially the magnetic behavior of the NPs. The obtained average crystallite size was 5.3–27.0 nm at 400 °C, 13.7–31.1 nm at 700 °C and 33.4–49.1 nm at 1100 °C, respectively. The $Zn_{0.6}Mn_{0.4}Fe_2O_4$ embedded in SiO_2 exhibited superparamagnetic-like behavior, whereas the unembedded $Zn_{0.6}Mn_{0.4}Fe_2O_4$ behaved like a high-quality ferrimagnet [2].

This Special Issue also includes the study of Dumitru et al. [11] on $Bi_2Cu(C_2O_4)_4 \cdot 0.25H_2O$ synthesis by thermolysis, followed by its integration within a CuBi/carbon nanofiber (CNF) paste electrode and its application in the electrochemical detection of amoxicillin (AMX) in aqueous solution. The obtained results indicated the potential use of a CuBi/CNF paste electrode for AMX detection in aqueous solutions. By adding a concentration step into the detection protocol, the selective and simultaneous detection of AMX in a multi-component matrix is also possible [11].

I am grateful to all the authors for their contributions, covering the most recent progress and new developments in the field of metal–ferrite nanocomposites, and hope that the published studies will pave the way for novel real-world applications of green nanomaterials.

Funding: This article received no external funding.

Data Availability Statement: Not applicable.

Acknowledgments: I am grateful to all the authors for submitting their studies to the present Special Issue, as well as to all reviewers for their helpful suggestions, which improved the manuscripts. I would also like to thank Keyco Li and the editorial staff of *Nanomaterials* for their excellent support during the development and publication of the Special Issue.

Conflicts of Interest: The author declares no conflict of interest.

References

1. Dippong, T.; Levei, E.A.; Cadar, O. Recent advances in synthesis and applications of MFe_2O_4 (M = Co, Cu, Mn, Ni, Zn) nanoparticles. *Nanomaterials* **2021**, *11*, 1560. [CrossRef]
2. Dippong, T.; Deac, I.G.; Cadar, O.; Levei, E.A. Effect of silica embedding on the structure, morphology and magnetic behavior of $(Zn_{0.6}Mn_{0.4}Fe_2O_4)_\delta/(SiO_2)_{(100-\delta)}$ nanoparticles. *Nanomaterials* **2021**, *11*, 23332. [CrossRef] [PubMed]
3. Chand, P.; Vaish, S.; Kumar, P. Structural, optical and dielectric properties of transition metal (MFe_2O_4; M = Co, Ni and Zn) nanoferrites. *Phys. B Condens. Matter* **2017**, *524*, 53–63. [CrossRef]
4. Asghar, K.; Qasim, M.; Das, D. Preparation and characterization of mesoporous magnetic $MnFe_2O_4@$ $mSiO_2$ nanocomposite for drug delivery application. *Mater. Today Proc.* **2020**, *26*, 87–93. [CrossRef]
5. Ozçelik, B.; Ozçelik, S.; Amaveda, H.; Santos, H.; Borrell, C.J.; Saez-Puche, R.; de la Fuente, G.F.; Angurel, L.A. High speed processing of $NiFe_2O_4$ spinel using laser furnance. *J. Mater.* **2020**, *6*, 661–670.
6. Kaur, H.; Singh, A.; Kumar, A.; Ahlawat, D.S. Structural, thermal and magnetic investigations of cobalt ferrite doped with Zn^{2+} and Cd^{2+} synthesized by auto combustion method. *J. Magn. Magn. Mater.* **2019**, *474*, 505–511. [CrossRef]
7. Naik, A.B.; Naik, P.P.; Hasolkar, S.S.; Naik, D. Structural, magnetic and electrical properties along with antifungal activity & adsorption ability of cobalt doped manganese ferrite nanoparticles synthesized using combustion route. *Ceram. Int.* **2020**, *46*, 21046–21055.
8. Nadeem, K.; Zev, F.; Azeem Abid, M.; Mumtaz, M.; ur Rehman, M.A. Effect of amorphous silica matrix on structural, magnetic, and dielectric properties of cobalt ferrite/silica nanocomposites. *J. Non-Cryst. Solids* **2014**, *400*, 45–50. [CrossRef]
9. Gharibshahian, M.; Mirzaee, O.; Nourbakhsh, M.S. Evaluation of superparamagnetic and biocompatible properties of mesoporous silica coated cobalt ferrite nanoparticles synthesized via microwave modified Pechini method. *J. Magn. Magn. Mater.* **2017**, *425*, 48–56. [CrossRef]
10. Duong, H.D.T.; Nguyen, D.T.; Kim, K.-S. Effects of process variables on properties of $CoFe_2O_4$ nanoparticles prepared by solvothermal process. *Nanomaterials* **2021**, *11*, 3056. [CrossRef] [PubMed]
11. Dumitru, V.; Negrea, S.; Păcurariu, C.; Surdu, A.; Ianculescu, A.; Pop, A.; Manea, F. $CuBi_2O_4$ Synthesis, characterization, and application in sensitive amperometric/voltammetric detection of amoxicillin in aqueous solutions. *Nanomaterials* **2021**, *11*, 740. [CrossRef] [PubMed]

nanomaterials

MDPI

Review

Recent Advances in Synthesis and Applications of MFe$_2$O$_4$ (M = Co, Cu, Mn, Ni, Zn) Nanoparticles

Thomas Dippong [1], Erika Andrea Levei [2] and Oana Cadar [2,*]

[1] Faculty of Science, Technical University of Cluj-Napoca, 430122 Baia Mare, Romania; thomas.dippong@cunbm.utcluj.ro

[2] National Institute for Research and Development for Optoelectronics INOE 2000, Research Institute for Analytical Instrumentation Subsidiary, 400293 Cluj-Napoca, Romania; erika.levei@icia.ro

* Correspondence: oana.cadar@icia.ro

Abstract: In the last decade, research on the synthesis and characterization of nanosized ferrites has highly increased and a wide range of new applications for these materials have been identified. The ability to tailor the structure, chemical, optical, magnetic, and electrical properties of ferrites by selecting the synthesis parameters further enhanced their widespread use. The paper reviews the synthesis methods and applications of MFe$_2$O$_4$ (M = Co, Cu, Mn, Ni, Zn) nanoparticles, with emphasis on the advantages and disadvantages of each synthesis route and main applications. Along with the conventional methods like sol-gel, thermal decomposition, combustion, co-precipitation, hydrothermal, and solid-state synthesis, several unconventional methods, like sonochemical, microwave assisted combustion, spray pyrolysis, spray drying, laser pyrolysis, microemulsion, reverse micelle, and biosynthesis, are also presented. MFe$_2$O$_4$ (M = Co, Cu, Mn, Ni, Zn) nanosized ferrites present good magnetic (high coercivity, high anisotropy, high Curie temperature, moderate saturation magnetization), electrical (high electrical resistance, low eddy current losses), mechanical (significant mechanical hardness), and chemical (chemical stability, rich redox chemistry) properties that make them suitable for potential applications in the field of magnetic and dielectric materials, photoluminescence, catalysis, photocatalysis, water decontamination, pigments, corrosion protection, sensors, antimicrobial agents, and biomedicine.

Keywords: transition metal; ferrites; magnetic nanoparticles; synthesis; applications

Citation: Dippong, T.; Levei, E.A.; Cadar, O. Recent Advances in Synthesis and Applications of MFe$_2$O$_4$ (M = Co, Cu, Mn, Ni, Zn) Nanoparticles. *Nanomaterials* **2021**, *11*, 1560. https://doi.org/10.3390/nano11061560

Academic Editor: Cesar De Julian Fernandez

Received: 9 April 2021
Accepted: 10 June 2021
Published: 13 June 2021

Publisher's Note: MDPI stays neutral with regard to jurisdictional claims in published maps and institutional affiliations.

1. Introduction

Nanocrystalline magnetic materials have attracted considerable interest due to their uniqueness and remarkable properties in various fields including physics, chemistry, biology, medicine, materials science, and engineering. Compared to their bulk counterparts, nanomaterials have particle size in the 1–100 nm range and a high surface to volume ratio that determines different or enhanced reactivity, thermal, mechanical, optical, electrical, and magnetic properties [1]. While in the case of bulk materials, the chemical composition is the main factor that determine their properties, in the case of nanomaterials, besides the chemical composition, the particle size and morphology determine most of their characteristics [1,2]. Moreover, these properties can be tunned based on the particle size and chemical composition [1,2].

Ferrites are important and interesting materials, from a practical as well as fundamental point of view. Among various ferrites, nanosized CoFe$_2$O$_4$, MnFe$_2$O$_4$, ZnFe$_2$O$_4$, NiFe$_2$O$_4$, and CuFe$_2$O$_4$ have attracted considerable attention due to their chemical and thermal stability and unique structural, optical, magnetic, electrical, and dielectric properties and wide potential technological applications in photoluminescence, photocatalysis, humidity-sensors, biosensors, catalysis, magnetic drug delivery, permanent magnets, magnetic refrigeration, magnetic liquids, microwave absorbers, water decontamination, ceramics pigment, corrosion protection, antimicrobial agents, and biomedicine (hyperthermia) [1,3–8].

3

$CoFe_2O_4$ has an inverse spinel structure with Co^{2+} ions mainly placed in octahedral (B) sites and Fe^{3+} ions almost equally distributed between tetrahedral (A) and octahedral (B) sites. It presents large coercivity (H_C), high magnetocrystalline anisotropy constant (K), high Curie temperature (T_C), moderate saturation magnetization (M_S), low remanent magnetization (M_R), excellent chemical and mechanical stability, rich redox chemistry, large magnetostrictive coefficient (λ), high electrical resistance, low eddy current losses, significant mechanical hardness, and low toxicity [7,9–13]. The synthesis of $CoFe_2O_4$ with different structures (i.e., nanoparticles, hollow mesoporous nanospheres, nanorods and three-dimensional ordered macroporous structure) have been reported [9,10,13]. Beside many advantages, $CoFe_2O_4$ owns poor electrical conductivity, poor cyclic stability, structural strain, and large volume expansion [9,10,14].

$ZnFe_2O_4$ has a normal spinel structure, where Zn^{2+} ions preferably occupy the tetrahedral (A) sites and Fe^{3+} ions the octahedral (B) sites [6,15,16]. The absence of Fe^{3+} in the A sites results in weak antiferromagnetic exchange interactions within Fe^{3+} in B sites, making $ZnFe_2O_4$ antiferromagnetic below 9 K [15–17]. The enhanced magnetization originates from the super-exchange interaction ascribed to the inversion of Fe^{3+} and Zn^{2+} ions in the tetrahedral (A) and octahedral (B) sites [18].

$NiFe_2O_4$ possesses an inverse spinel structure, where the tetrahedral (A) sites are occupied by Fe^{3+} ions, while the octahedral (B) sites by Fe^{3+} and Ni^{2+} ions, the Fe^{3+} ions being easily distributed between the A and B sites [19,20]. $NiFe_2O_4$ is one of the most versatile and technologically important soft ferrite materials due to its low electrical conductivity, high electrochemical stability, catalytic behavior, abundance in nature, low K and H_C values, high M_S, paramagnetic, superparamagnetic or ferrimagnetic behavior depending on the particle size and shape, low eddy current loss and conductivity, high electrical resistance, and electrochemical stability [4–6,11,21].

$CuFe_2O_4$ has an inverse spinel structure with 8 Cu^{2+} ions on octahedral sites and 16 Fe^{3+} ions equally distributed between the tetrahedral (A) and octahedral (B) sites [22,23]. The Cu^{2+} ion displays the transition from tetragonal phase, technologically less important (at low temperature) to cubic phase, which is relatively more useful (at high temperature) [24,25]. Moreover, magnetic and electrical properties of $CuFe_2O_4$ vary significantly with the change of cation distribution [25]. $CuFe_2O_4$ exhibits ferrimagnetism, high thermal stability, high resistance to corrosion, excellent catalytic properties, and sufficient band gap, acting as an efficient photocatalyst [22–24,26,27].

Manganese ferrite ($MnFe_2O_4$) has a partially inverse spinel structure, fine structural, magnetic and electrical properties, with 20% of Mn^{2+} ions located at octahedral (B) sites and 80% located at tetrahedral sites [3,8]. $MnFe_2O_4$ nanoparticles (NPs) has attracted attention in biomedicine due of its good biocompatibility, controllable size, high magnetization value, superparamagnetic nature and ability to be monitored by an external magnetic field. Moreover, $MnFe_2O_4$ is an inorganic heat-resistant, non-corrosive, environmentally friendly, non-toxic, high shock resistant [28], reusable adsorbent, often employed for adsorption and desorption processes [29] and magnetic drug targeting/delivery [3].

The surface coating of magnetic NPs is required in order to generate non-toxic, biocompatible and water dispersible NPs, along with drug targeting ability. The most used surface coating materials are poly(vinylalcohol), poly(N—isopropylacrylamide), polyethylene glycol (PEG), chitosan, Au, ZrO_2, and SiO_2 [3]. Of these, due to the excellent drug loading capacity, stability, non-toxicity, biocompatibility, and water dispersibility, the mesoporous SiO_2 is a multipurpose candidate for the development of nanocarriers, as it enhances the stabilization of ferrite NPs in water, improves the chemical stability, and minimizes the agglomeration of NPs, without influencing their magnetic and dielectric properties [30]. Non-magnetic SiO_2 can easily promote conjugation with many functional groups, thus allowing selective and specific coupling and labeling of biotargets. Moreover, SiO_2 coatings may change the surface properties of magnetic NPs and offer a chemically inert layer, which is particularly useful in biological systems [30,31].

By searching in the Web of Science Core Collection the keyword "M ferrite", where M = cobalt, copper, nickel, manganese, and zinc, we observed that Co ferrite attracted the attention, with the number of publications on Co-ferrite exceeding by far those published on the other ferrites. Between the studied ferrite, Co-ferrite was the topic of 4276 papers, followed by Zn-ferrite (3073), Ni-ferrite (2432), Mn-ferrite (895), and Cu-ferrite (880). For every ferrite, the number of papers started to grow exponentially in the last 20 years (Figure 1), the highest increasing rate being observed for the Co-ferrite. The increasing interest in these ferrites may be attributed both to the development of new equipment that allowed the ferrites characterization and the increase of the demand for materials with special properties for a wide range of applications.

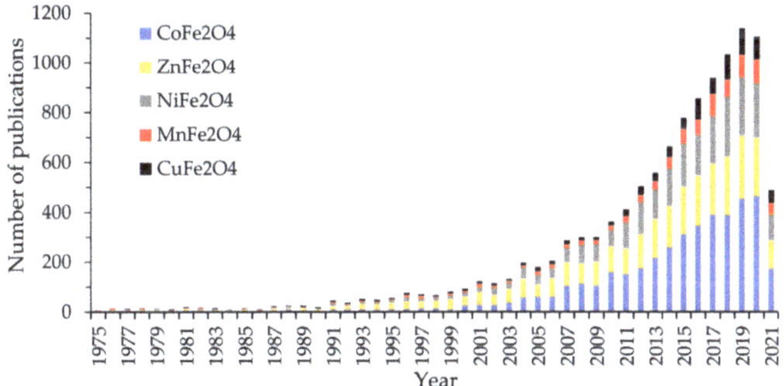

Figure 1. Papers having M-ferrite (M = Co, Cu, Ni, Mn, Zn) in the topic published in Web of Science Core Collection between 1975–June 2021.

In the last few years, the number of papers that review data on nanosized ferrites with a focus on different ferrites, different synthesis methods, or different applications significantly increased. The Web of Science Core Collection have indexed 58 publication that contains the words "ferrite" and "review" in their title of which 25 were published between 2019 and June 2021.

Zate et al. [32] reviews the mechanical, chemical, spray and electrospining methods used for the synthesis of different ferrites with magnetic properties. Vedrtnam et al. [33] reviews the properties, classification, synthesis, and characterization of hexagonal and spinel ferrites with a focus on the main four synthesis routes (sol-gel, hydrothermal, co-precipitation and solid-state), magnetic properties, and characterization of the ferrites. Vinosha et al. [34] review the recent advances of synthesis, magnetic properties, and water treatment applications of cobalt ferrite, while Masunga et al. [35] reviews the recent advances in copper ferrite synthesis, magnetic properties and application in water treatment. Kumar et al. [36] review magnetic nano ferrites and their composites used in the treatment of pollutants from waste waters, emphasizing pros and cons of several synthetic pathways, the adsorption mechanism, and ferrite regeneration, while Kefeni and Mamba [37] review the photocatalytic application of spinel ferrite NPs in pollutant degradation with emphasis on the possible recovery and reuse of NPs. Kharisov et al. [38] review the use of cobalt, nickel, copper, and zinc ferrites and their doped derivates as catalysts in organic processes, while Dalawai et al. [39] overview the spinel-type ferrite thick films together with preparation strategies and sensors, microwave, magnetic, and advanced applications. Kefeni et al. [40] review the ferrite's applications in electronic devices, such as sensors and biosensors, microwave devices, energy storage, electromagnetic interference shielding, and high-density recording media together with the advantages and drawbacks of most important ferrite NPs synthesis methods.

Our previous works reported the thermal, structural, morphological, and magnetic characterization of ferrites (MFe_2O_4, M = Co, Mn, Zn, Ni and Cu) and doped ferrites with different divalent transition metals, produced by sol-gel synthesis [41–46]. Furthermore, since the physical, chemical, magnetic, electrical, and optical properties can be tailored by the dopant type and content, we also reviewed the potential applications of $CoFe_2O_4$ and divalent transition metal-doped cobalt ferrites ($M_xCo_{1-x}Fe_2O_4$, M = Zn, Cu, Mn, Ni, and Cd) [47]. In addition, this paper aims to deepen our previous studies and review a number of topics including the various synthesis methods of $CoFe_2O_4$, $MnFe_2O_4$, $ZnFe_2O_4$, $NiFe_2O_4$, and $CuFe_2O_4$ NPs together with their advantages and disadvantages, and the most important applications in conventional and modern technologies. The review not only summarizes the existing literature from theoretical and methodological points of view, but also synthetizes it from a new perspective, representing a significant and useful contribution to subsequent research.

2. Synthesis Methods

The chemical and physical properties of transition metal nanoferrites (MFe_2O_4; M = Co, Ni, Zn, Mn, Cu) are dependent on the synthesis method and conditions. Thus, the selection of an appropriate synthesis route plays a crucial role in tailoring the properties and obtaining high quality nanoferrites [1]. Nanosized ferrites can be synthesized by various techniques (Figure 2) and the possibility to obtain almost any solid solution of nanoferrites unlocks the way to tailor their properties for a wide range of applications [22].

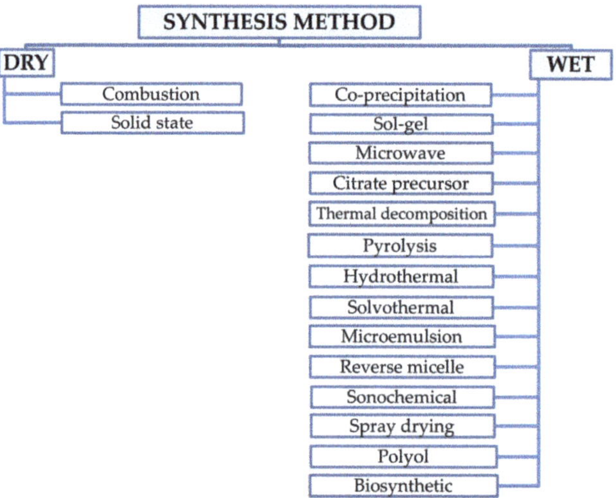

Figure 2. Classification of nanoferrite synthesis methods.

There are different classification approaches of the ferrite synthesis methods: (i) physical, chemical, and biological, based on the type of processes that take place in the synthesis methods, (ii) dry and wet methods, based on the presence or absence of a solution, and (iii) conventional and non-conventional, based on their novelty. Despite several classification methods, it is difficult to univocally group the synthesis methods, as sometimes different processes are coupled to obtain NPs with specific characteristics.

Many efforts have been made to tailor the size, shape, particle size distribution, surface area, composition, structure, and properties of the ferrite NPs by employing different synthesis methods, or changing the synthesis parameters, such as the annealing temperature and duration, concentration of reactants, pH value, stirring speed, doping additives, etc. [48,49]. Numerous physical and chemical methods have been used for the synthesis of ferrites [1,19,25]. Some synthesis methods are high-energy consuming, complex procedures, demanding a high processing temperature and long reaction time to complete the crystallization, and the use of reduction agents with negative effects on the environment [6,13]. The importance of these factors is different in every method, making the achievement of reproducibility in desired properties difficult [50]. Another issue that appears in the majority of the synthesis methods is the agglomeration of NPs after production, which limits the control of size, shape and function [51]. The wet-chemical synthesis has multiple advantages. However, single phase ferrite can be obtained only after annealing at high temperatures and is accompanied by particle growth, aggregation, and coarseness of NPs [14,20].

2.1. Dry synthesis Methods

2.1.1. Combustion Method

Combustion method is simple, fast, and inexpensive, as it does not involve intermediate decomposition or calcinations steps. This method exploits an exothermic, usually very rapid, and self-sustaining chemical reaction between metal salts and an organic fuel (glycine, urea, citric acid, sucrose, hydrazine, and polyvinyl alcohol) that act as a reducing agent [14,52]. The powder characteristics (i.e., crystallite size, surface area) are governed by the enthalpy or flame temperature produced during the combustion, which is dependent on the type of fuel and fuel-to-oxidant ratio. Beside the key role of fuel in the morphology of the NPs it also determines the phase formation [53]. An important aspect is that the heat necessary to sustain the chemical reaction is provided by the reaction itself and not by an external source [8]. Nitrates are the favorite metal precursors because of their high solubility in water and easy combustion following their mixing with a suitable fuel [7,54]. Ammonium nitrate is used as an extra oxidant in the combustion reaction, producing the expansion of the microstructures and eventually the increase of NPs surface area, without changing the proportion of the other participants [7]. The combustion method is a very popular method for the synthesis of ceramic materials, composites, and ferrimagnetic nanomaterials, due to its efficiency, short preparation time, use of relatively inexpensive precursors and simple equipment, good control of stoichiometry, desired particle size distribution, formation of high-purity products, and stabilization of metastable phases. The main disadvantages are high combustion temperature and low production yields [6,11,13,14,55]. The combustion method is a good choice for the preparation of high quality $CoFe_2O_4$, $NiFe_2O_4$ and $MnFe_2O_4$ NPs. By varying the nitrates to fuel ratio, the size, M_S and H_C values are tailored [20,56,57]. Nanocrystalline $NiFe_2O_4$ was successfully prepared by mixing of metal nitrates and citrate with the formation of a colloidal solution (sol), followed by the continuous heating of xerogel, the auto combustion process until the formation of a loose powder, and the annealing of the powder at 700 °C [21]. This method requires less time and produces pure and homogeneous NPs, without any type of waste product, but involves a high temperature [7,58,59]. The high purity of materials prepared by the combustion process is attributed to the removal of unwanted impurities as volatile species at high temperatures [6].

2.1.2. Solid State Synthesis

Solid state synthesis produces polycrystalline ferrite nanomaterials from solid reagents at high temperature [20]. This method assumes the grinding and mixing of metal nitrates or sulphates with NaOH or NaCl in an agate mortar or mills for short times. After the removal of NaCl by washing, the powders are dried at 80 °C for two hours and then annealed at 700 °C for two hours [60]. $CoFe_2O_4$ NPs have been synthesized by a low-temperature solid state process using various salts (sulphate, acetate, nitrate and chloride). By using chloride salts, the obtained nanomaterials have smaller particle size than by using other salts [61]. Furthermore, in comparison to other synthesis techniques, the low temperature solid state process requires neither complex apparatus, nor solvent or solution, and is a convenient, environment-friendly, low-cost, time-saving and low energy consumption process [62]. The $CoFe_2O_4$ NPs prepared using solid state technique display high M_S values, as well as low M_R and low H_C values [63]. The major advantages are that the synthesis procedure is completed under atmospheric pressure, which is economic and feasible, as well as the use of non-expensive and toxic solvents and cost-effective raw materials [58]. Mechanical milling is a simple, low-cost, solid state and non-equilibrium process in which the final product have a very fine, typically nanocrystalline or amorphous structure. Generally, the final product has a nanosized structure with enhanced properties and performance compared to the bulk material [15]. The main advantages of this method are simplicity, low cost and ability to produce large volumes [63]. Mechanical milling of appropriate metal oxides for 12 h and subsequent sintering at 600 °C for 2 h were applied to prepare $ZnFe_2O_4$ and $NiFe_2O_4$ ferrites [15]. $CoFe_2O_4$ NPs were obtained by ball milling of Fe_2O_3 and Co_3O_4 powders in stoichiometric amounts using high-energy vibratory mill for 8 h, followed by calcination at 900 °C, for 12 h [64]. $CoFe_2O_4$ NPs of 10 nm size were also prepared by the precipitation of hydroxide/oxidhydroxide, followed by mechanical milling at lower speeds, and subsequent heat treatment for a short time. NaCl was added before milling to avoid agglomeration [62]. $NiFe_2O_4$ crystallites were prepared by high energy ball milling. The increase of milling time led to a significant increase in the $NiFe_2O_4$ formation and to a progressive decrease in its particle size and lattice parameter [20]. $MnFe_2O_4$ NPs (4–8 nm) were prepared by solid state ball-milling and calcinations (300–400 °C) of nitrate precursors and citric acid [65].

2.2. Wet Synthesis Method

2.2.1. Co-Precipitation Method

Co-precipitation method is one of the most widely used methods due to its high yield and simplicity in producing high purity ultrafine magnetic nanostructured ferrites [58,66]. The co-precipitation method is a very simple and cost-effective method that allows an easy control of particle size and composition, requires low temperature and leads to materials with high crystallinity, homogeneity and good textural properties [1,58]. The major drawbacks of co-precipitation methods are extensive agglomeration, poor crystallinity and particle size distribution, and the necessity of pH control [57,67].

In this case, homogeneous solutions are formed by dissolving inorganic salts (chloride, sulfate, nitrate) in water or solvents. After pH adjustment in the range of 7–12 under continuous stirring, the precipitate is collected by filtration or centrifugation, washed, and dried. The pH change rate causes the particles aggregation and crystal growth [58]. The most common way to synthesize nanostructured ferrites by chemical co-precipitation method is using Co^{2+} and Fe^{3+} salts in the presence of a strong base [51]. By adjusting the experimental parameters (i.e. reaction temperature and time, reagents feeding rate and concentration, pH, drying temperature, etc.), the size, shape, and magnetic properties of the nanostructures may be controlled [68].

$CoFe_2O_4$ NPs (2–47 nm) were synthesized from metal chloride salts using different concentration of aqueous NaOH and NH_4OH solution using a reaction time of 2 h and variable reaction temperature (20–100 °C). To achieve the precipitation reaction, the aqueous metal chloride solutions were slowly added into the preheated boiling aqueous alkaline solution [68]. The crystallite size depends on the reaction temperature, time, concentration of the base solution, and pH [68]. Another study reported the preparation of $CoFe_2O_4$ NPs (2–14 nm) by controlling the co-precipitation temperature of Co^{2+} and Fe^{3+} ions in alkaline solution from 20 to 80 °C [49].

The co-precipitation of $CoFe_2O_4$ nanocrystals by mixing Fe^{2+}, Co^{2+} and NaOH in the presence of oxidizing agent such as KNO_3 to convert Fe^{2+} into Fe^{3+} was reported by Chia et al. [66] and Senapati et al. [69]. However, due to the coexistence of both Fe^{2+} and Fe^{3+} ions, besides $CoFe_2O_4$, secondary phase such as magnetite (Fe_3O_4) is also formed [66]. Moreover, the H_C values of particles obtained by this procedure are not very high and depends on the amount of surfactant used, that favor a stable colloidal dispersion of the NPs, as well as on the annealing temperature [69].

A detailed investigation on the effect of precursors, solvents, precipitating agent, and heat treatments on the particle size, particle size distribution, morphology, and chemical composition of $CoFe_2O_4$ NPs synthesized at various liquid phase was reported by Prabhakaran et al. [68]. In some cases, in order to facilitate the solubilization of metal salts at low temperatures or to prevent oxidation and particle agglomeration, some surfactants such as oleic acid are added to the solution before precipitation [1].

The temperature may influence the size and stability of the NPs [50]. The particle size growth in the co-precipitation method is influenced by the difference between local and surface temperature, as the formation of spinel nano-ferrites is usually an exothermic reaction, with the latent heat being released at the surface [70]. At low temperatures particles tend to aggregate, the particle size increasing with the decrease of temperature. High purity and no aggregation of the NPs were reported at high temperatures due to directed crystallization [50].

Reverse co-precipitation is similar to conventional co-precipitation, except that instead of adding a precipitant into the precursor ions solution, the precursor ions are added into the precipitant solution. Thereby, the precipitant is in a supersaturated state, ensuring the complete precipitation. The resultant NPs have smaller particle-size than those obtained by traditional co-precipitation [71]. The synthesis of $CoFe_2O_4$ NPs by reverse co-precipitation was reported by Huixia et al. [58].

2.2.2. Sol-Gel Method

The sol-gel method is a low temperature process based on hydrolysis and condensation reactions of metal precursors (salts or alkoxides), leading to the formation of a three-dimensional inorganic network [72]. Sol results from the conversion of monomers into a colloidal solution, while gel is obtained after the solvent evaporation by joining together particles into a network [58]. The sol-gel method is a simple, low cost, and environmentally friendly method to prepare nanocomposites (NCs) as it allows good control of the microstructure, particle size, dispersion, structure, and chemical composition by carefully monitoring the preparation parameters [58,73,74]. The sol-gel method has been used to prepare very fine, highly dense, homogenous, and single-phase ferrite NPs. Compared to other conventional methods, the sol-gel displays a good stoichiometric control and also allows the production of ferrites at relatively low temperature [20]. The obtained nanomaterials may be formed either as films or as colloidal powders [58]. The main disadvantage is the limited efficiency and long duration of the synthesis process [32,48,58]. $NiFe_2O_4$ nanostructures with an average particle size of 27 nm were synthesized by sol-gel method using glycolic acid as a chelating agent [20]. In the case of $CoFe_2O_4@SiO_2$, the SiO_2 network protects the NPs and minimizes the surface roughness and spin disorder. The H_C values at room temperature (RT) for $CoFe_2O_4@SiO_2$ were much higher than that of unembedded $CoFe_2O_4$ [73]. The most commonly used reagents are metal nitrates as metals source, 1,2-ethanediol, 1,3-propanediol or 1,4-butanediol as chelators, tetraethyl orthosilicate (TEOS) as matrix precursor, ethanol as solvent and HNO_3 [32]. By annealing, the NPs agglomerate and form larger particles. Thus, further research on the influence of thermal energy released during heat treatment on the undesired growth of particle size is needed [70].

The Pechini method is an alternative approach to sol-gel synthesis and consists in the complexing of cations with hydroxycarboxylic acids (usually citric acid or ethylenediaminetetraacetate) in an aqueous-organic medium. The obtained chelates are cross-linked and transformed into polymers using polyalcohol (ethylenediol, polyvinyl alcohol) through polyesterification. By heating, the viscous liquid is dried and forms a gel. By annealing the gel, the organic part is removed, resulting in reactive oxides, which further form the ferrites [75–77]. The microwave-assisted Pechini method combines the advantages of microwave and Pechini methods, such as low temperature and time of process, accurate control of stoichiometry, uniform mixing of various components on molecular level, low price of precursor and equipment, and capability of industrial scale-up [31,75]. However, the main drawback of this method is the long and energy-wasting thermal treatments in order to remove the large amounts of organic precursors [75]. The microwave-assisted sol-gel Pechini method is a faster, energy-saving procedure for obtaining single-phase nanopowders of high purity [75].

Combining the sol-gel and auto-combustion methods results in a simple and inexpensive preparation method for high purity, homogenous nanopowders at low annealing temperature [56,61]. The method consists in dissolving nitrate salts in water, addition of the organic complexing agent (citric acid), adjustment of the solution to pH 7, heating at

70 °C to form the sol, and then at 110 °C to remove the residual water and form the gel, and initiation of the autocatalytic self-combustion process of nitrate-fuel gel [78]. The citric acid and glucose are two of the most used fuels due to their strong complexing ability and low ignification temperature (200–250 °C) [78]. The synthesis of $CoFe_2O_4$ NPs in the size range of 11–40 nm by the sol-gel auto combustion method and annealing at different temperatures (800–1000 °C) was also reported [78].

The organic precursor method is similar to the Pechini method and involves the preparation of aqueous solution containing cations, chelation of cations using carboxylic acid, followed by heating of the solution until precursor formation and annealing of the precursors. The carboxylic acid acts both as complexing agent and as organic rich fuel. Magnetic $CoFe_2O_4$ NPs (38.0–92.6 nm) were produced by the organic precursor method using oxalic and tartaric acids as precursors [79].

2.2.3. Microwave-Assisted Combustion Method

The microwave-assisted combustion method applies microwave radiations to the reaction solution to synthesize NPs. Microwave radiation is absorbed and converted to thermal energy, the heat being generated inside a material, unlike the conventional heating methods where the heat is transferred from the outside. This heating allows for a considerable reduction of processing time and energy [13,54,80]. The main advantages of this method compared to other synthesis methods are simplicity, homogeneous nucleation, rapid reaction time, high production rates, environmentally friendly, excellent reproducibility, low energy cost, easy handling, and control of parameters. In the microwave-assisted combustion process, the reagents are mixed at the molecular level, due to the interaction of microwaves, offering an excellent control of stoichiometry, purity, homogeneity, and morphology [53,58,59]. However, the method is expensive mainly due to the expensive fuels such as urea, glycine, L-alanine, carbohydrazide, or citric acid used to promote and control the combustion process in accordance with the propellant chemistry principles, and inappropriate for scale up and reaction monitoring [13,53,80–82].

In the case of the microwave hydrothermal method, the necessary heat is generated by microwaves, which have the advantage of very fast heating to a certain depth, allowing the generation of homogenous nanomaterials with fine particle size distribution [52,53]. Zn, Fe, Mn, Cu, Ni, and Co nitrates were dissolved in deionized water, maintaining a pH equal to 9.4. The mixture was sealed with tetrafluorometoxil (TFM) and placed in a microwave oven at 160 °C for 30 min. The resulting wet mixture was dehydrated, followed by adding polyvinyl alcohol (PVA) as a binder and sintering for 30 min [58]. This method provides faster heating, is more economical, and produces very fine, uniform NPs [59].

2.2.4. Citrate Precursor Method

The citrate precursor method is a wet chemical process that involves the mixing of aqueous solutions of precursor salts (metal nitrates) with an aqueous solution of citric acid, followed by heating at about 80 °C and annealing at 700 °C [83,84]. During this process, the precursors are thermally decomposed into ferrite powders [83]. The advantage of this method is the high reactivity, low reaction time, low synthesis temperature, homogenous distributions of ions, and low cost over other chemical methods [83]. To reduce the particle agglomeration, the dispersion of NPs in SiO_2 matrix is a commonly used. The synthesis of $CoFe_2O_4@SiO_2$ NPs by citrate method using TEOS, ethanol, Fe and Co nitrates, citric acid, and ethylene glycol was successfully reported by Garcia [83] and Varma [62], with the latter reporting 50 nm size and M_S values in the range of 4–25 emu/g.

2.2.5. Thermal Decomposition Method

The thermal decomposition method is based on the heating of reactants (metallic precursors) at different temperatures. To control the nucleation and growth of the NPs and consequently improve the physical properties of the material, such as crystallite size, porosity, and specific surface, surfactants (oleylamine, oleic acid) are added in the decomposition step. The surfactants act as a protective envelope coating the NPs, limiting the coalescence, and leading to the improvement of the physical properties of the material, such as crystallite size, porosity, and specific surface. These parameters allow the control of the magnetic properties [85].

The method is relatively simple, low-cost, takes place at low reaction temperature, is environment friendly, produces highly monodispersed particles with a narrow size distribution, and does not produces byproducts. The main drawback is the removal of surfactants simultaneously with the particle size controlling [50,86]. The use of thermal treatment to synthesize ferrite NPs is very limited, although this method requires reduced time as compared to other synthesis techniques [87]. The thermal decomposition method, reflux temperature, and time play an important role in controlling size distribution and aggregation of $CoFe_2O_4$ NPs, which will significantly influence the magnetic properties such as permeability and coercivity of the products. The reaction rate depends on the concentration of metal precursors, surfactants, reducing agents and temperature [88]. One example of thermal decomposition is the dissolution of metal nitrates in deionized water, followed by their dispersion in a solution of polyvinylpyrrolidone (PVP) at 70 °C, followed by a thermal treatment at 600 °C for four hours [87]. Asghar et al. [3] reported the synthesis of $MnFe_2O_4$ by dissolving Fe and Mn acetates (2:1, molar ratio) in dibenzyl ether, in the presence of 1,2-dodecanediol, oleic acid, and oleylamine, followed by a thermal treatment at 250 and 350 °C, respectively. Peddis et al. reported the synthesis of crystalline $CoFe_2O_4$ NPs coated by oleic acid and organized in a self-assembling arrangement with narrow size distribution (~5 nm) by high thermal decomposition using Fe(III) acetylacetonate, Co(II) acetylacetonate, 1,2-hexadecanediol, oleylamine, oleic acid, and phenyl ether as solvents, heated to 200 °C for 30 min and to 265 °C for 30 min [89]. Following the thermal treatment method, in aqueous solution containing metal nitrates and polyvinyl pyrrolidone, followed by grinding and calcination at temperatures ranging between 723 and 873 K, $MnFe_2O_4$ NPs (12–22 nm) were produced [86].

2.2.6. Pyrolysis

Spray Pyrolysis Method

The spray pyrolysis method consists in converting the reagent mixture into aerosol droplets, solvent evaporation, solute condensation, and drying followed by thermolysis of the particles at high temperature [50]. This method allows the control of the particle formation environment by dividing the solution into droplets [90]. Generally, it is suitable for the synthesis of mixed metal ferrites as it ensures complete stoichiometry retention on the droplet scale and provides a highly homogeneous distribution of the components [90]. By controlling the type of thermolysis reaction, the type of precursors, the gaseous carrier, deposition time, and substrate temperature, this method allows the synthesis of a broad range of hollow or porous particles with potential applications in thermal insulation or catalyst support [90,91].

Non-agglomerated particles smaller than 10 nm are achieved by using soluble inert additives (NaCl, H_3BO_3) to the reaction mixture. Furthermore, the inert components allow subsequent thermal treating of the obtained powder without significant particle size growth [89]. The method's advantages include obtaining predictably sized, finely dispersed particles of variable shape and composition, high production rate, scalability, and reduced time [50,89]. The disadvantages of the method are the high costs and possibility to obtain aggregated particles [50]. $Zn_xFe_{3-x}O_4$ and $CoFe_2O_4$ thin films were produced by spray pyrolysis technique, by spraying aqueous solution of metal nitrates at a temperature of 300 °C. After deposition, film was naturally cooled [92]. $CoFe_2O_4$ thin film is prepared by the spray pyrolysis technique due to its easy handling, simple experimental setup, and cost effectiveness as large surfaces of thin film can be produced [92].

Laser Pyrolysis

Laser pyrolysis is another method for the synthesis of nanosized ferrites. The gaseous phase precursors are transported by a carrier gas introduced into a reaction chamber where a high-power laser beam (2400 W) generates elevated local temperatures, which trigger the nucleation and growth of NPs that are further collected by a filter [93]. Liquid reactants are introduced into the reaction chamber reactants as vapors or microscale droplets [94]. This technique allows the production of high purity nanomaterials with small and narrow particle size distribution in one step. The method is versatile and flexible, and allows the control of operational parameters, such as laser power, type of gases, gas flow rate, and concentration of the precursors [94]. Moreover, it can be easily scaled-up to industrial production. The disadvantages of the method include the high cost of equipment used and the need for a precursor with absorption band at 10.6 μm, or addition of C_2H_4 or NH_3, which leads to further contamination with carbon or nitrogen [93,94]. The synthesis of $ZnFe_2O_4$ nanopowders by laser pyrolysis using Fe and Zn nitrate solutions as precursors, ethylene as sensitizer gas and air as carrier gas was reported. The obtained nanopowders were tested as negative electrode materials for Li-ion batteries [94].

2.2.7. Hydrothermal Method

The hydrothermal method is used for the preparation of ferrite NPs at large scale owing to its high yields [58,59]. The method consists in the NPs formation by mixing the aqueous divalent metal (Ni, Co, Mn, Zn, Mn) acetate solutions with iron nitrate and a carbon-based nano template at alkaline pH and its dispersion around 200 °C in an autoclave. The obtained precipitate is washed, separated by centrifugation, and annealed around 500 °C to eliminate the carbon-based template [95]. In order to control the crystal structure, particle size, and morphology, surfactants are often employed in solution process [96]. Ethylenediamine and citric acid may also be used in the synthesis. This method requires neither sophisticated processing nor high processing temperature and allows the selection of the NPs properties by using different temperatures, pressure, reaction times, and nano templates [59,97].

Advantages of the hydrothermal method are good nucleation control, production of low particle size with narrow particle size distribution, controlled morphology at a high reaction rate and different temperature and pressure levels, high yield, excellent stoichiometry, and simplicity [67].

The production of magnetic nanosized $CoFe_2O_4$ by hydrothermal method without subsequent calcination processes was reported [98]. However, the influence of the synthesis time on the morphology and the particle size of $CoFe_2O_4$ NPs is still under discussion [96]. $CoFe_2O_4$ nanorods using cetyltrimethylammonium bromide (CTAB) as the surfactant were also synthesized by the hydrothermal method [99]. Porous anodic aluminum oxide is also used as template in the synthesis of one-dimensional $CoFe_2O_4$ nanostructures, however the costs of the synthesis are high while the yields are low [99]. The decrease of the reaction medium pH favors the $CoFe_2O_4$ crystallite size decrease. However, pure $CoFe_2O_4$ at pH < 10 was not obtained independently of the temperature and reaction time [100]. $CoFe_2O_4$ nanoplatelets and NPs were also prepared by hydrothermal reaction from $FeCl_3$, cobalt dodecyl sulfate and NaOH aqueous solution, in strictly controlled synthesis environment [101]. $CoFe_2O_4$ NPs were also synthesized by the supercritical hydrothermal method at different temperatures of 25–390 °C for two hours [58,59,68,97].

$NiFe_2O_4$ nanocrystals were synthesized from $FeCl_3 \cdot 6H_2O$ and $NiCl_2$ $6H_2O$, using cetyltrimethylammonium bromide (CTAB) as the surfactant and NH_3 and NaOH as hydrolyzing agents [101]. Moreover, $ZnFe_2O_4$ were synthesized using metallic Zn sheet and $FeCl_2$ as reactants in ammonia solutions [101]. Hydrothermal synthesis of MFe_2O_4 (M = Ni, Co, Mn, Zn) using metal acetylacetonate and aloe vera extracts provided high-yield nanosized ferrite with well crystalline structure using high calcination temperature [101].

2.2.8. Solvothermal Method

The solvothermal method has emerged as a promising technique to synthesize monodisperse spherical magnetic microspheres with high surface area, magnetic saturation and good dispersion in liquid media [102]. $CoFe_2O_4$ magnetic NPs in the size range 2–15 nm were prepared using a non-aqueous solvothermal method by dissolving Co(III) acetylacetonate and Fe(III) acetylacetonate in acetophenone, followed by solvothermal treatment between

120 and 200 °C for 22 h in a Parr Acid Digestion Bomb autoclave [103]. The solvothermal method was successfully applied to synthetize $CoFe_2O_4$, $NiFe_2O_4$, $MnFe_2O_4$, and $ZnFe_2O_4$ using sodium or ammonium acetate, polyethyleneglicol, urea, and oleylamine at around 200 °C [102]. Size-controlled $NiFe_2O_4$ NPs were successfully synthesized via a simple solvothermal method using ethylene glycol as solvent and sodium acetate as electrostatic stabilization, by adjusting the experimental parameters (time, initial concentration of reactants, amount of protective reagents, and the type of acetates) [20]. The main advantages of the method are high performance in biological applications by obtaining magnetic NPs with smooth surfaces, narrow size distributions, large surface areas, and high magnetic saturation in order to provide maximum signal and good dispersion in liquid media [49].

2.2.9. Microemulsion Method

The microemulsion method consists in the mixing of microemulsions containing reactants, the formation of micro droplets, followed by the trapping of fine aqueous micro droplets inside surfactant molecule assembles. This procedure results in a locking up effect that limits the growth dynamics, particle nucleation, and agglomeration during the nanoparticle formation. The major advantage is the obtaining of monodispersed NPs with various morphologies, while among the major disadvantages can be listed the low efficiency and difficulty to scale up [69]. $CoFe_2O_4$ NPs lower than 50 nm, with high H_C and M_S were synthetized by a simple micro-emulsion synthesis route using water-in-oil emulsions consisting of water, cetyltrimethyl ammonium bromide as surfactant, n-butanol as co-surfactant, and n-octane as oil phase [104]. $CoFe_2O_4$ NPs of 4–25 nm were prepared using normal micelle microemulsion methods [105].

2.2.10. Reverse Micelle Method

The reverse micelle method is based on the formation of water-in-oil emulsions in which the water to surfactant ratio controls the size of water pools within which aqueous chemical syntheses take place [106]. The method consists in mixing the metal salt precursors and the precipitating agent and the formation of microemulsion droplets, which act as a nano-reactor, in which nanometric size precipitates are formed following the coalescence and droplet collision of reactants [107]. This method allows an excellent control of the particle size, size distribution, chemical stoichiometry, and cation occupancy and is suitable for room temperature reactions such as the precipitation of oxide NPs [106].

Metallic chlorides and NaOH were dissolved by ultrasonication in two different microemulsion systems prepared by mixing water, sodium dodecyl sulfate, 1-butanol and n-hexane. The two microemulsions were mixed together till metallic hydroxides were precipitated, the obtained precipitates were filtered, washed, dried and annealed at 400 °C [107]. The nature of surfactant, water to surfactant ratio and pH value of solution strongly influences the particle size of ferrites, which in turn affects their properties [101]. For the synthesis of $CoFe_2O_4$ a surfactant system formed by mixing sodium dioctylsulfosuccinate with a 2,2,4-trimethylpentane oil phase was used [106]. $CoFe_2O_4$ magnetic NPs were synthesized by reverse micelle methods using a micro emulsion consisting of three independent phases, (petroleum oil, pyridine and water) as a template to control the size of the NPs [108]. $MnFe_2O_4$ NPs with particle size of approximately 8–25 nm were prepared using reverse micelle microemulsion methods [105]. Using the reverse micelle microemulsion method, a wide range of spinel ferrite nanoparticle cores can easily be coated with a silica shell [105]. Such a method increases the potential for development of tunable magnetic silica for magnetoelectronic and biomedical applications [105].

2.2.11. Sonochemical Method

The sonochemical method is considered one of the most promising procedures to obtain ferrite NPs. The sonochemistry arises due to acoustic cavitation phenomenon consisting in the formation, growth and collapse of bubbles in liquid medium. The extreme reaction conditions (i.e., high temperature (5000 K), pressure (20 MPa) and cooling rate

(1010 K/s)) lead to numerous unique properties of the produced particles [109]. Highly crystalline, monodisperse $CoFe_2O_4$ NPs with uniform spherical shape and high M_S values were synthetized in a one-step, surfactantless, sonochemical process. Additionally, the synthesis time was low (about 70 min) and no subsequent annealing was necessary [109]. The advantages of this approach over the conventional methods include simplicity, low cost, safety, environment friendly, uniform size distributions, high surface area, fast reaction time, and good phase purity [109]. Disadvantages comprise a very small concentration of prepared NPs and particle agglomeration [50].

2.2.12. Spray Drying Method

The spray drying method is used for the large-scale production of $ZnFe_2O_4$ and $MnFe_2O_4$. The method is based on the spraying of the ferrite slurry droplets into a hot air vertical evaporating tube. The main drawbacks are the inflation defects visible as large voids inside the granule and the development of capillary stress due to the rapid water evaporation. This stress may lead to undesirable diffusion and segregation phenomena, which reduce the compositional and morphological homogeneity of NPs [110]. As the defects may resist annealing, to preserve the requested particle size and avoid the occurrence of defects, the spray droplets may be frozen using liquid nitrogen and freeze-dried. During water sublimation, there are no undesirable capillary stresses that generate hard unbreakable aggregates and result in voids that lead to a rigid porous product of loose and non-agglomerated particles [110].

2.2.13. Polyol Method

The polyol method consists in the reduction of metallic oxide or metallic complexes by a high boiling point solvent that acts as a solvent as well as a reducing agent of the metallic ions under reflux conditions. The method allows controlling particle growth and preventing the particles agglomeration by the suspension of precursors into liquid polyol (ethylene glycol, diethylene glycol and triethylene glycol), followed by thermal treatment up to boiling point of polyol. The major advantage of this method is the obtaining of uniform size soft and hard magnetic NPs, while the main disadvantages are the high temperature and long time [111,112]. The polyol method has been used for synthesis of many oxides and NCs materials [73]. To produce $CoFe_2O_4$, $FeCl_3 \cdot 6H_2O$ and $CoCl_2 \cdot 4H_2O$ were individually dissolved in glycol, mixed together under continuous stirring. To the homogenous solution, water and sodium acetate were added, and the pH value of the solution was adjusted (8, 10, and 12) by adding ammonia solution. Under reflux conditions, the mixture was heated to the boiling temperature of glycol, while the formed NPs were separated by centrifugation [111]. The synthesis of $NiFe_2O_4$ NCs starting from tetraethoxysilane without, or with modifiers as formamide, citric acid, and PVA were also reported [4].

2.2.14. Biosynthetic Method

The biosynthetic method uses plant extracts as a simple and viable alternative to chemicals during the synthesis process. The use of microorganisms, enzymes, and plant extracts has been proposed as a possible eco-friendly alternative to chemicals of physical synthesis. Furthermore, the biosynthetic route is very simple and provides high-yield, crystalline nanomaterials with adequate properties [13]. Aloe vera (*Aloe barbadensis mill.*) plant-extract may be used as a bio-reducing agent in the preparation of ferrites, providing a simple, efficient and green alternative route for the synthesis of nanomaterials. A possible explanation for this could be the presence of long chain polysaccharides in the Aloe vera plant extract that allows the homogeneous distribution of ferrite [13]. The major advantages consist in the selectivity and precision of NPs formation, cost-effectiveness, as well as the use of an eco-friendly reducing agent and non-hazardous gelling agent for stabilizing the nanostructures. Disadvantages are the difficult size and properties controlling, high temperature required to convert the precursors in crystalline materials and the formation of polydisperse, surface capped or unpurified NPs, in addition to poor reproducibility [13,50].

The use of Ni and Fe nitrates and freshly extracted egg white (ovalbumin) in an aqueous medium, results in the formation of $NiFe_2O_4$ NPs with ferrimagnetic behavior, M_S values in the range of 26.4–42.5 emu/g for the applied field of 10 kOe, at room temperature [20].

In conclusion, although different synthetic routes have been adjusted, the key challenge in the synthesis field remains the obtaining of size- and phase-controlled synthesis, with good reproducibility. However, reproducibility is difficult to achieve in the synthesis methods where partial mixing of reactants and undesired reactions take place, affecting the properties of NPs. Moreover, the heat and mechanical treatments can alter crystal structure, affecting the photocatalytic activity. The combustion method is simple and quick, but the rate of organic fuel to nitrates can highly affect the size and magnetic properties. The solid-state method at higher temperatures produced micron-sized $CoFe_2O_4$ particles, while the citrate precursor method at lower temperatures produced NPs. To overcome these challenges in an ideal, controlled reproducible synthesis, the following steps should be achieved: (i) direct active and complete mixing of reactants, (ii) automation, and (iii) enabled reaction parameters controlled precisely. Each synthesis method has its own pros and cons, and selection depends on many factors, but amongst all methods of synthesis, the sol-gel and chemical co-precipitation techniques stand to be favorable routes to synthesize homogeneous, highly pure, and narrow size distribution nanoferrites.

3. Applications

The most frequent applications of MFe_2O_4 (M = Co, Cu, Mn, Ni, Zn) are presented in Figure 3. Considering the cation type and synthesis route, the structure of the ferrite and its particle size and shape changes and thus also its properties, which further lead to different applications.

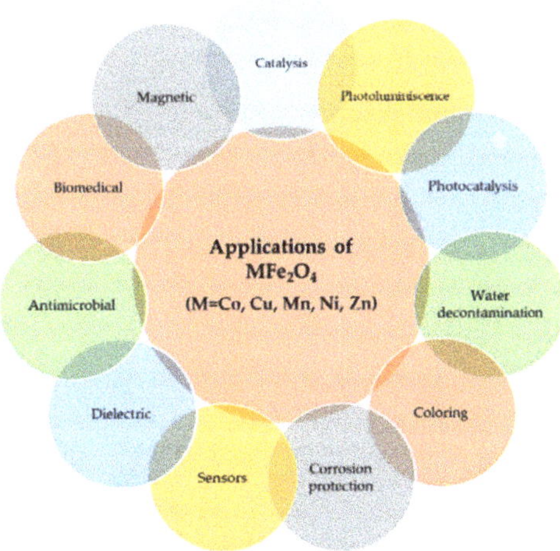

Figure 3. Application of nanosized ferrites.

3.1. Magnetic Applications

The dependence of the magnetization on the grain size is due to the variations of exchange interaction between tetrahedral and octahedral sites and [20,113]. For their application in high-density magnetic recording, the magnetic particles should have nanosize to prevent the exchange interaction between adjacent grains, and thus reducing the media noise in the materials. The particles must also possess high H_C values to obtain high storage density [98]. The magnetic properties (M_R, M_S, and H_C) of spinel ferrites depend upon the composition, particle size, crystal structure, and cationic distribution between octahedral and tetrahedral sites. Moreover, they may exhibit antiferromagnetic, ferrimagnetic and paramagnetic behavior [101,114].

The increase of H_C values may also occur due to a combination of spin disorder, spin canting effect, and improved surface barrier potential in surface layers [87,115]. In the case of magnetic NPs, the presence of a large number of atoms at the surface due to large surface-to-volume ratio results in interesting and superior properties in comparison to corresponding bulk materials. Moreover, the low number of coordination surface atoms relative to interior atoms of the nanoparticle leads various surface effects such as spin canting, spin disorder and existence of a magnetically dead layer [115].

High coercivity materials are known as hard materials, while low coercivity materials are known as soft materials [113]. The soft materials are used for inductor cores, transformers, and microwave devices, while hard materials are used for permanent magnets. Generally, the H_C value is low in soft ferrites and the magnetization can be tailored for cutting-edge applications in electronic engineering such as microwave components, high-frequency inductors and transformer cores [58]. Furthermore, due to the high H_C values of hard ferrites, it is not so easy to magnetize, and hence it is used as enduring magnets with applications in washing machine, refrigerator, communication systems, microwave absorbing systems, loudspeaker, TV, switch-mode power supplies, DC-DC converters, and high-frequency applications [58].

$CoFe_2O_4$ is a ferrimagnetic ceramic with excellent properties, such as high K, H_C, and T_C values, moderate M_S values, and large magnetostrictive coefficient value [42,116]. The high M_S and M_R values can be explained on the basis of Néel's theory and cation distribution at tetrahedral (A) and octahedral (B) sites [11,25,117]. Generally, in single-phase samples the magnetic characteristics are highly influenced by the material microstructure such as shape and size of crystals, residual stress and crystal defects [51]. Moreover, the synthesis temperature plays a key role in controlling particle size of $CoFe_2O_4$ with a significantly influence on its magnetic properties [66]. As expected, M_S values increase with the $CoFe_2O_4$ content and temperature, due to the increase of the crystallinity degree as well as to the average size of the magnetic $CoFe_2O_4$ NPs [14,42,99]. Due to the high values of K and high value of spin-orbit coupling constant of Co^{2+} ions, $CoFe_2O_4$ displays the highest magnetostriction (i.e., up to 200 ppm) among oxide-based materials. The increment of H_C value is mainly attributed to the increasing particle size as a consequence of the coarsening of nanocrystals and diminished crystal defects at high temperature [98]. In the case of $CoFe_2O_4$ embedded in SiO_2 matrix ($CoFe_2O_4@SiO_2$), the reduction of M_S and M_R values is attributed to the reduced amount of magnetic material per gram of nanocomposite [105]. Moreover, the increase of annealing temperature leads to the decrease of K and H_C values [74]. In contrast, the H_C value of $CoFe_2O_4@SiO_2$ NPs does not show any change after coating, while the H_C value of $MnFe_2O_4$ NPs decreases by 10% after coating [105]. In the case of $CoFe_2O_4$ obtained by the citrate gel method, higher M_S values were obtained for samples synthesized at high temperatures than the samples prepared at low temperatures, suggesting that the M_S value rises with increase in particle size, while the M_R decrease with the increase of particle size [62]. Peixoto et al. [43] produced $CoFe_2O_4$ NPs embedded in a SiO_2 matrix via sol–gel method and evaluated the magnetic properties at 5 K and 100–200 K, revealing a superparamagnetic behavior.

The M_S value of 27.09 emu/g found for $CuFe_2O_4$ obtained by solid state chemistry (particle size of 36 nm) was similar to the M_S value of 33 emu/g for $CuFe_2O_4$ NPs of ~32 nm obtained by other synthesis method. However, the H_C of 112 Oe found for $CuFe_2O_4$ obtained by solid state chemistry (particle size 36 nm) is greater compared to other syntheses, as a consequence of the presence of some impurities (CuO). Further, an M_R value of 4 emu/g was reported for $CuFe_2O_4$ of particle size ~8 nm synthesized by the sol-gel method [113]. The H_C of $CoFe_2O_4$ NPs show a maximum H_C value at a particle size of about 25 nm, while in the case of $CuFe_2O_4$ NPs obtained by sol-gel, co-precipitation, solid-state reaction, thermal decomposition, and solution auto-combustion methods, the H_C decreases linearly with the decreasing size, indicating that the particles are in single-domain area and the surface spins dominate the magnetization decreases with size reduction [57]. The H_C and

M_S values of $CuFe_2O_4$ were higher in the case of the combustion method and lower in the case of the precipitation method [57].

The increase of M_R and M_S values of $MnFe_2O_4$ can be ascribed to the metal cations distribution at octahedral and tetrahedral sites of crystal lattice structure, meaning to the tendency of Mn^{2+} ions to be positioned at octahedral (B-site) [118]. $MnFe_2O_4$ NPs and $MnFe_2O_4@SiO_2$ nanocomposite exhibit superparamagnetic behavior attributed to the ultra-small size of $MnFe_2O_4$ NPs and are widely used for drug delivery applications. The low M_S value of $MnFe_2O_4@SiO_2$ nanocomposite is due to the presence of non-magnetic SiO_2 matrix [3]. The H_C of $MnFe_2O_4@SiO_2$ NPs slightly decreases compared to that of native magnetic NPs. No similar behavior was observed for $CoFe_2O_4@SiO_2$ probably due to the larger contribution of the surface anisotropy to the total anisotropy of $MnFe_2O_4$ NPs [105]. The M_S value of 52.4 emu/g obtained in the $MnFe_2O_4$ (diameter of ~15.9 nm) is lower than values 67.0 emu/g for synthesized $MnFe_2O_4$ crystallites using a triethanolamine-assisted route under mild conditions (diameter of ~1 μm), but higher than the value of ~48.6 emu/g for the polymer-pyrolysis route $MnFe_2O_4$ NPs (diameters of ~9 nm) [101].

$ZnFe_2O_4$ displays antiferromagnetic behavior when the temperature is below the Néel temperature and diamagnetic, superparamagnetic, or ferrimagnetic behavior when particle sizes are at nanometer scale [119]. The superparamagnetic behavior is due to the increased disorder of magnetic moments orientation in the various sites when the ratio surface/volume increases. The M_S value of $ZnFe_2O_4$ strongly depends on various factors, such as the synthesis route and its conditions, type of the precursors, and annealing treatments [63]. The M_S value of 7.06 emu/g for $ZnFe_2O_4$ (diameter of ~17.9 nm) is lower than values of 54.6 emu/g for hydrothermal-synthesized $ZnFe_2O_4$ ultrafine particles (crystallite size of ~300 nm) [101]. However, the paramagnetic behavior of $ZnFe_2O_4$ was remarked upon in bulk [46] and NPs, due to the increase of an atypical Fe^{3+} cation distribution in the tetrahedral coordination sites upon nanosizing [47,101].

In the crystal structure of $NiFe_2O_4$, the Ni^{2+} occupies the octahedral site with Fe^{3+} cations distributed equally at tetrahedral and octahedral sites, while their antiparallel spins produce a net magnetic moment of $2\mu_B$ owing to Ni^{2+} ions at octahedral site [87]. Due to high crystallinity and uniform morphology, $NiFe_2O_4$ ferrites show high M_S and low H_C values [119]. The magnetic properties of $NiFe_2O_4$ NPs are attributed to the cumulative effect of various factors, such as super-exchange interaction, magneto crystalline anisotropy, canting effect, and dipolar interactions on the NP's surface. Besides, the variation of H_C value with particle size can be explained by the domain structure, critical size and anisotropy of the crystal [20]. The large $NiFe_2O_4$ is due to the scattering in direction of anisotropy field due to inhomogeneous broadening, at high temperature, it has the tendency to make magnetic moment isotropic causes the decrease of H_C value [21]. The M_S value of 31.9 emu/g found for $NiFe_2O_4$ (particle size of ~8.2 nm) is close to the values of 34.5 emu/g for $NiFe_2O_4$ NPs (crystallite size of ~68 nm) obtained by egg-white solution route, but lower than the theoretical M_S of 50 emu/g calculated using Neel's sublattice theory and the reported value of 56 emu/g for the bulk sample [96]. The M_S value of 55.3 emu/g obtained in the $CoFe_2O_4$ (diameter of ~8.5 nm) is lower than values of 80 emu/g for bulk $CoFe_2O_4$ and ~65 emu/g for the $CoFe_2O_4$ NPs with crystallite size of ~40 nm synthetized by aerosol route, and higher than the values of 30 emu/g for hydrothermal-synthesized $CoFe_2O_4$ NPs (diameter of ~30 nm).

The synthesis of nanocrystalline spinel ferrite of type MFe_2O_4 (e.g., M = Mn, Co, Ni, Cu, Zn) has great relevance to modern technological applications in several industrial and biological fields, including magnetic recording media and magnetic fluids for the storage and/or retrieval of information, magnetic resonance imaging enhancement, and magnetically guided drug delivery [101]. Furthermore, the magnetic NPs showed promising results in the field of medicine and healthcare treatment owing to their biocompatibility, low toxicity, and ability to be handled by the application of a magnetic field [50].

Table 1 presents the main magnetic parameters (M_S, M_R, H_C, K), particle diameter, and synthesis method of the studied ferrites.

Table 1. Average crystallites size (D), saturation magnetization (M_S), remanent magnetisation (M_R), coercivity (H_C) and anisotropy constant (K) of $CoFe_2O_4$, $CuFe_2O_4$, $MnFe_2O_4$, $NiFe_2O_4$, $ZnFe_2O_4$, produced by different synthesis methods; RT-room temperature.

Ferrite	Synthesis Method	Temp.	D_{XRD} (nm)	M_S (emu/g)	M_R (emu/g)	H_C (kOe)	$K{\cdot}10^3$ (erg/cm^3)	Ref.
CoFe$_2$O$_4$	sol-gel	RT	28.0	26.3	11.9	1.440	2.378	[46]
	sol-gel auto-combustion	RT	36.0	89.0	20.0	0.650	-	[11]
	sol-gel auto-combustion	RT	52.0	65.6	71.9	1.117	-	[78]
	solid-state	RT	80.0	77.2	-	0.525	-	[120]
	co-precipitation	RT	17.3	78.6	15.8	0.778	-	[121]
	co-precipitation	RT	37.0	72.6	34.5	1.060	-	[117]
	co-precipitation	RT	11.70	58.4	12.45	0.286	-	[108]
	combustion	RT	52.0	52.6	20.8	1.274	-	[122]
	hydrothermal	RT	17.3	63.4	23.1	0.831	0.450	[123]
	hydrothermal	RT	8.5	55.3	-	0.074	-	[101]
	hydrothermal	RT	16.3	58.3	-	1.029	-	[121]
	normal micelles	RT	5.58	12.6	0.17	0.0237	-	[108]
	reverse micelles	RT	7.62	29.4	0.84	0.0252	-	[108]
	reverse micelle	RT	6.00	34.4	8.50	0.500	-	[124]
	polyol	RT	14.0	57.0	21.6	1.123	-	[125]
	thermal decomposition	RT	13.0	68.0	-	0.560	-	[88]
	thermal decomposition	RT	82.8	48.38	11.72	0.64	-	[126]
CuFe$_2$O$_4$	sol-gel	RT	60.0	14.5	3.03	0.018	0.163	[46]
	sol-gel	RT	8.00	16.5	4.00	0.450	-	[113]
	sol-gel auto-combustion	RT	18.6	78.9	35.9	0.705	0.569	[127]
	solid-state	RT	38.0	27.1	1.41	0.112	-	[113]
	co-precipitation	RT	31.0	22.9	1.12	0.114	-	[113]
	co-precipitation	RT	52.0	28.1	8.52	0.189	-	[117]
	combustion	RT	22.3	14.0	0.17	0.006	-	[118]
	auto-combustion	RT	22.0	18.9	1.51	0.140	-	[113]
	thermal decomposition	RT	53.0	29.0	0.91	0.102	-	[113]
	thermal decomposition	RT	10–250	0.90	-	0.600	-	[128]
	sonochemical	RT	40.0	21.5	12.6	0.235	-	[24]

Table 1. *Cont.*

Ferrite	Synthesis Method	Temp.	D_{XRD} (nm)	M_S (emu/g)	M_R (emu/g)	H_C (kOe)	$K \cdot 10^3$ (erg/cm^3)	Ref.
MnFe$_2$O$_4$	sol-gel	RT	49.0	28.5	14.8	0.119	3.392	[46]
	co-precipitation	RT	6.10	13.5	1.24	0.046	-	[129]
	hydrothermal	RT	22.0	65.5	3.86	0.054	0.018	[130]
	hydrothermal	RT	15.9	52.4	-	43.9	-	[101]
	hydrothermal	RT	25.4	67.3	-	0.173	-	[124]
	polyol	RT	7.10	51.9	-	-	-	[130]
	sonochemical	RT	25.5	59.4	3.41	0.024	1.53	[131]
	microwave combustion	RT	27.9	61.0	11.4	0.064	-	[53]
	thermal decomposition	RT	61.0	56.0	10.0	0.08	-	[132]
NiFe$_2$O$_4$	sol-gel	RT	23.0	20.1	6.82	0.061	0.770	[46]
	sol-gel	RT	70.0	37.3	-	0.321	-	[133]
	sol-gel auto-combustion	RT	58.0	50.0	7.00	0.050	-	[11]
	co-precipitation	RT	17.3	43.9	16.6	0.051	-	[121]
	combustion	RT	25.0	30.2	4.00	0.159	-	[122]
	microwave combustion	RT	18.5	37.9	2.57	0.016	-	[134]
	hydrothermal	RT	8.20	31.9	-	0.007	-	[101]
	thermal decomposition	RT	25.0	36.5	10.6	0.263	-	[87]
	thermal decomposition	RT	79.0	43.60	17.63	0.645	-	[135]
	thermal decomposition	RT	10.7	36.8	-	-	-	[136]
ZnFe$_2$O$_4$	sol-gel	RT	49.0	10.8	1.67	0.015	0.102	[46]
	sol-gel auto-combustion	RT	54.0	5.31	0.08	0.113	-	[78]
	solid-state	RT	80.0	77.27	-	0.525	-	[120]
	co-precipitation	RT	4.80	11.9	0.01	0.004	-	[129]
	microwave combustion	RT	37.5	2.60	0.01	0.007	-	[53]
	microwave combustion	RT	21.1	3.85	0.51	0.010	-	[134]
	hydrothermal	RT	15.9	52.4	-	0.044	-	[101]
	solvothermal	RT	80.0	77.0	-	0.090	-	[102]
	reverse micelle	RT	8.30	4.90	0.01	0.010	-	[107]
	thermal decomposition	RT	30.0	12.8	-	-	-	[137]

The magnetostrictive materials are solids in which the application of a magnetic field results in strong strain and deformation as a consequence of the strong coupling of magnetic moments with crystal lattice [138]. The large magnetostriction in a wide range of temperature is very useful when a mechanical coupling is required in multiferroic composite systems [139]. The electronic and optical properties of magnetostrictive solid materials are more sensitive to magnetic field comparing to solids with weak magnetostriction [138]. Magnetostrictive smart materials are generally used to design sensors, actuators, sonar transducers, motors, etc. In this regard, there is an increasing interest on metal oxide based magnetostrictive materials such as $CoFe_2O_4$ due to various practical advantages over alloy based magnetostrictive materials, i.e., the high magnetostriction and strain-field derivative with low saturation field of $CoFe_2O_4$ results in potential for sonar detector and force sensor applications and vibration components in high frequency ultrasonic transducer [140–142]. The magnetostriction value of $CoFe_2O_4$ prepared using micron-size powders without magnetic annealing decreases from 130 ppm to 200 ppm [142]. The magnetic annealing may increase the magnetostriction and strain derivative values due to the uniaxial anisotropic distribution of magnetic domains, while various magnetic or non-magnetic ions could decrease the magnetostriction and increase the strain derivative values [142]. The application of a magnetic field to magnetostrictive ferrimagnetic $CoFe_2O_4$ leads to strong strain and deformation of the crystal lattice and to change of spectrum-strain-magneto-optics, while high magnetoreflection is remarked on in the case of great magnetostriction [138].

Beside the excellent soft magnetic properties, Fe-based soft magnetic composites have unique properties such as high electric resistivity and low eddy loss, making them good electrical insulation coatings without any notable decrease in magnetic properties [143]. H_C value confirms the magnetic properties of hard magnetic or soft magnetic materials. $CoFe_2O_4$ obtained using co-precipitation method and low heat treatment have M_S of 21.74 emu/g, M_R of 2.37 emu/g, and H_C of 556.57 Oe [144]. The in-situ oxidation is an effective and novel method for manufacturing soft magnetic composites with low core loss for applications in medium and high frequency fields. The nano-$ZnFe_2O_4$ layer effectively improves the magnetic properties of soft magnetic composites [143]. Soft magnetic spinel ferrite properties probably occur due to the interactions within metal oxide and particular vacancy oxygen ions in their cubic spinel structure [144].

3.2. Photoluminescent Applications

Room temperature photoluminiscence is one of the important properties of mixed spinels as $CoFe_2O_4$, $NiFe_2O_4$ and $ZnFe_2O_4$ nanostructures. The photoluminescence spectrum offers information regarding the surface oxygen vacancies and defects, as well as the efficiency of the charge carrier trapping, immigration and transfer [17]. The broad visible band emission was ascribed to the charge transfer among Fe^{3+} at octahedral sites, M^{2+} (M = Co, Ni and Zn) at both tetrahedral and octahedral sites, and its surrounding O^{2-} ions [1]. The blue emission peak at 460 nm is attributed to the Fe^{3+} transition in the ferrite sites, while the main peak at 418 nm resulted from the free electrons trapped at the oxygen vacancies. The intensity of every peak diminished with the increasing Ni content, as a consequence of the increasing band gap which reduces the electron-hole recombination ratio [145]. The violet emissions resulted from the radiating defects associated to the interface traps within the grain boundaries. The decrease in the luminescence intensity of Co^{3+} from $CoFe_2O_4$ with the increase of doping fraction suggests a lower electron-hole recombination ratio [81]. Photoluminescence measurements showed a value around 2.13 eV for the energy gap of nanocrystalline $ZnFe_2O_4$ [63].

3.3. Catalytic Applications

Spinel ferrites have been extensively used as heterogeneous catalysts as they can be simply recovered from the reaction mixture by filtration or in the presence of an external magnetic field and recycled numerous times, making the process more economically and environmentally feasible [53]. Heterogeneous catalytic nanomaterials play a central role in the selective protection of functional groups in order to be economically valuable and environmentally friendly. The catalytic property of MFe_2O_4 (M = Cu, Ni, Co, Zn) spinel revealed that $CoFe_2O_4$ catalyst displayed the best performance using benzaldehyde with 63% conversion and 93% selectivity, while $CuFe_2O_4$ catalyst using H_2O_2 oxidant showed 57.3% conversion and 89.5% selectivity [53]. The particle size, surface area, morphology, and chemical composition considerable influence the catalytic activity of materials [134]. In addition, the catalytic property of nanocrystalline spinel ferrites depends on the distribution of cations among the tetrahedral (A) and octahedral (B) sites. As catalyst, $CoFe_2O_4$ displays the best performance compared to other ferrites [13]. The catalytic activity of $CoFe_2O_4$ was successfully reported for the synthesis of arylidene barbituric acid derivatives. The advantages of this method were short reaction time, high yields, simple and economic work-up procedure, high turnover frequency, chemoselectivity, environmental sustainability, and green conditions [38]. Large-size magnetic $CoFe_2O_4$ NPs were successfully used as catalyst for the oxidation of various alkenes in the presence of tert-butylhydroperoxide. The catalyst from the medium was easily separated using an external magnet and the catalyst was recycled several times without important loss of activity. $CoFe_2O_4$ exhibited high catalytic activity and selectivity in methanol decomposition to CO and H_2 [38]. Moreover, non-nanosized $CoFe_2O_4$ displayed catalytic applications, such as the conversion of CO to CO_2. In the case of the catalytic oxidation of benzyl alcohol to produce a mixture of benzaldehyde, benzoic acid, and benzyl benzoate, the $CoFe_2O_4$ NPs have been proven to be highly selective [13]. $CoFe_2O_4$ NPs can be also used as catalyst for Knoevenagel condensation reactions between aromatic aldehydes and ethylcyanoacetate, in mild conditions. The catalyst could be easily recovered from the reaction mixture using an external magnet and can be re-used four times, without a significant loss of activity [69].

$NiFe_2O_4$ is an effective catalyst in several industrial processes. Its catalytic activity related to O_2 yield and evolution rate under ambient reaction conditions, is comparable to that of Ir, Ru or Co-based materials [119]. Due to its importance for both industry and domestic safety, the catalytic combustion of CH_4 has attracted considerable interest, because of its higher energy conversion efficiency and low emissions of environmental pollutants. Therefore, the development of a high activity catalyst with sensitivity for CH_4 detection is of great interest [19]. In this regard, the catalytic performance of $NiFe_2O_4$ for CH_4 combustion, but also for alkylation and oxidation reactions has been reported [19]. $NiFe_2O_4$ was successfully used also as a catalyst in photocatalytic water oxidation using $[Ru(bpy)_3]^{2+}$ as a photosensitizer and $S_2O_8{}^{2-}$ as a sacrificial oxidant [119].

$CuFe_2O_4$ has been used in industrial processes as a catalyst in both organic and inorganic reactions [146]. The main advantages of $CuFe_2O_4$ NPs are simple work-up and low-cost procedure, mild reaction conditions, reusable catalyst, high yield, short reaction times, no isomerization during the reaction [38]. The use of $CuFe_2O_4$ NPs in organic catalysis, such as the reaction of substituted aromatic aldehydes, ethyl acetoacetate, and ammonium acetate, at room temperature was also reported. In all cases, the nano-catalyst can be easily recovered and re-used [9,38]. $CuFe_2O_4$ NPs were also employed as reusable heterogeneous initiator in the synthesis of 1,4-dihydropyridines, of α-aminonitriles and conversion of CO to CO_2 [38,146]. The successful use of $CuFe_2O_4$ NPs as catalyst for the ligand free N-arylation of N-heterocycles and for the cross-coupling of aryl halides with diphenyl diselenide to produce diaryl selenides was also reported [147,148]. The $CuFe_2O_4$ NPs also proved to be efficient catalysts for a simple, one pot, green, and efficient synthesis method for substituted benzoxazoles by redox condensation [149].

$ZnFe_2O_4$ spinel acts as a wide band gap semiconductor with application in hydrogen generation by light-activated water splitting and in photocatalytic removal of harmful organic molecules [85]. Non-nanosized $ZnFe_2O_4$ have been used as catalysts in the oxidative conversion of methane, oxidative coupling of methane, and in methanol decomposition to CO and H_2O [38]. Additionally, the coupling between the catalyst and magnetic material is a new approach to enhance the catalytic performance of a catalyst [150].

3.4. Photocatalytic Applications

Photocatalysts are important materials that support the use of solar energy in oxidation and reduction processes, with applications in various areas, such as the removal of water and air contaminants, odor control, bacterial inactivation, water splitting to produce hydrogen, the inactivation of cancer cells, etc. [151]. Nowadays, the usage of photocatalysis is the preferred treatment methods for dyes removal, as alongside irradiation of light on a semiconductor, the produced electron-hole pairs are used also for the oxidation and reduction process [81]. The degradation of dyes occurs due to the formation of active radicals during the photocatalytic reaction [22,145].

Only few materials are capable of both photo-oxidation and photo-reduction, namely to complete the decomposition of harmful organic compounds and, concomitantly, to absorb visible light efficiently. The low crystallite size of ferrites leads to a large surface area and more reaction sites, increasing the photocatalytic activity [151]. The type of photocatalyst, crystallinity, size of NPs, accessibility of the active surface to pollutants, and diffusion resistance of organic pollutants are the most important characteristics that enhance the photocatalytic properties. Thus, small particle sizes with high crystallinity are of great interest due to their higher specific surface area and active sites that favor the photocatalytic activity [37]. Due to their spinel crystal structure and a band gap capable to absorb visible light, ferrites are suitable as photocatalysts for the degradation of a wide range of contaminants [151].

$CoFe_2O_4$ was used as photocatalyst for the degradation of organic dyes (i.e., methylene blue (MB) and rhodamine B (RhB)), its chemical stability and the narrow bandgap (1.1–2.3 eV) making it active under visible light [81,145]. $CuFe_2O_4$ photocatalyst was found to be more effective than $ZnFe_2O_4$ and $NiFe_2O_4$ in decomposing hazardous dye compounds, probably due to lower crystallite size, large surface area, small band gap and occurrence of pores that trap oxygen molecules and produce large number of oxidizing species and OH• radicals. The photocatalyst was recovered using a magnet and reused in four consecutive cycles under visible light [22]. $CuFe_2O_4$ "oversized" nanostructures displayed improved photocatalytic activity in the conversion of benzene under Xe lamp irradiation. Moreover, $CuFe_2O_4$ powders exhibited a good catalytic efficiency (~60%) at 200 °C and pH = 12 due to the high surface area (~120 m^2/g) [38]. The photocatalytic ozonation of dyes with $CuFe_2O_4$ NPs prepared by co-precipitation method was reported to successfully decolorize and degrade textile dyes (Reactive Red 198 and 120) without using high pressure of oxygen or heating [152].

The photocatalytic activity of $ZnFe_2O_4$ NPs relies on the surface properties and defects. $ZnFe_2O_4$ was tested for solar energy conversion and photochemical hydrogen production due to its relatively narrow band gap energy (1.9 eV), excellent visible-light response, good photochemical stability and low-cost [17]. The use of $ZnFe_2O_4$ NPs for the photocatalytic degradation of 4-chlorophenol and the dependence of the degradation efficiency on their particle size and morphology was also reported [134]. $ZnFe_2O_4$ was found to be an appropriate visible-light-driven catalyst, which completely degraded the RhB dye in short time, due to the small particle size and narrow size distribution [107]. The sol–gel/precipitation hybrid synthesis method of magnetic $NiFe_2O_4@TiO_2$ and its use as photocatalyst for the production of hydrogen from aqueous systems was reported by Kim et al. [153].

3.5. Water Decontamination

In the last years, the industrial wastewater management is one of the main challenges in developed countries. Wastewaters resulted from the textile industry contain many non-biodegradable organic dyes mixed with different contaminants in a variety of ranges. Untreated effluents negatively affect not only humans and animals, but also the flora and fauna [81]. RhB is a synthetic, highly toxic, water soluble, organic dye, widely used as colorant in many industries and frequently found in wastewaters. Various techniques, such as ozonation, electrochemical method, and Fenton process, have been employed for the treatment of RhB containing water. The use of $CoFe_2O_4$ for wastewater treatment is based on its high adsorption capacity, magnetic properties that allow the NPs separation using an external magnetic field, and on its low energy band gap that enables the photocatalytic degradation of RhB under visible light irradiation, [37,81]. Moreover, $CuFe_2O_4$ is considered a prospective material for the water decontamination by photoelectrochemical processes. In combination with a suitable adsorbent, $CuFe_2O_4$ can be applied as a cost-effective adsorbent for removal of MB from water and wastewater [146]. The atrazine degradation by a three-dimensional electrochemical process using $CuFe_2O_4$ NPs as electrodes and catalyst for the activation of persulfate for were also reported [149]. Further, $ZnFe_2O_4$ can be effectively used as magnetically recyclable material for the removal of chemical contaminants (i.e., dyes), as well as biological contaminants from water/industrial wastewater treatment [154].

3.6. Coloring

Ceramic pigments are metal transition oxides with high thermal and chemical stability, high tinting strength when dispersed and fired with glazes or ceramic matrices, high refractive index, acid and alkali resistance, and low abrasive strength. The color of each pigment is obtained by adding chromophore agents (usually transition metals) in an inert oxide matrix [155].

$CoFe_2O_4$ is a black pigment widely used in the ceramic industry. The color performance depends on the coating crystallization degree, a larger number of crystals in the glass resulting in a brighter color [156]. The $ZnFe_2O_4$ spinels are thermally stable, insoluble, and resistant to aggressive media, exhibit a high covering power and a reasonable cost. $ZnFe_2O_4$ spinel pigments improve the mechanical strength of the binder by a chemical reaction producing cationic soaps, resulting in low solubility and a tendency towards saponification in contact with a corrosive environment [157]. The color of $ZnFe_2O_4$ depends on the annealing temperatures and particle size as it causes a reduction of the total reflecting surface of the powder. At high annealing temperatures, the system suffers a notable modification due to the disappearance of defects (as oxygen vacancies), leading to less distorted tetrahedral and octahedral sites, and consequently better-defined colors [158]. The possibility to select the desired cations in the spinel lattice (Zn^{2+} in A sites and Fe^{3+} in B sites for $ZnFe_2O_4$) can be used to change or to create new properties of the pigment dispersed in an organic binder [157].

3.7. Corrosion Protection

Metal corrosion is caused by an electrochemical process, generally promoted by Cl^- ions, which dissolve Fe oxides [159]. The corrosion of a metallic substrate may be constrained by the transformation of ferrous (Fe^{2+}) ions to ferric (Fe^{3+}) ions, which act as inorganic corrosion inhibitors in the electrochemical process, leading to the formation of a passivation layer, which will overcome the cathodic and/or anodic reaction [157].

Organic coatings act as barrier or mechanical protection against corrosive environments [157]. Nanocontainers can release corrosion inhibitors and subsequently protect the metal substrate from corrosion when incorporated into an organic coating [160]. The $CoFe_2O_4@SiO_2$ nanopigment enhances the corrosion protection performance of the coating better than $CoFe_2O_4$, by its appropriate dispersion in the coating and by filing and blocking the free cavities and all electrolyte pathways in the coating. Besides, it releases inhibitive species (Co^{2+} cations) in the scratched area and restricts corrosion inhibition at the coating/metal interface. The incorporation of $CoFe_2O_4@SiO_2$ nanopigment into epoxy coatings results in a noticeable improvement of the coating corrosion protection [160]. Furthermore, the corrosion protection of pigments, such as diatomite, talc, wollastonite, and kaolin, was considerably improved using the surface treatment with a layer of $ZnFe_2O_4$ on the particle surface. Low amounts (16–20 wt.%) of $ZnFe_2O_4$ used for the surface treatment also enhance the corrosion protection [157].

3.8. Sensors

Sensors are devices used for signaling changes that appear in a specific material in a certain environment [40]. Sensors based on ferrite NPs are highly sensitive, have low detection limits, and a high signal to noise ratio [40]. One of the most frequent uses of sensors is for detecting changes of humidity. The monitoring of humidity is frequently used in industrial and domestic environments in order to assure human comfort and the appropriate storage of various goods, as well as the optimal functioning condition for industrial process and various instruments [161]. Generally, the humidity-sensing phenomenon is based on the surface effect of water vapor–solid interaction. The ceramic type humidity sensors based on metal oxides displayed superior advantages comparing to polymer films in terms of physical stability mechanical strength, thermal capability, and resistance to chemical attack, making them promising materials for electrochemical humidity sensor applications. By water adsorption, the electrical properties (i.e., capacitance, resistance or electrolytic conduction) of the ceramic surfaces change. In this regard, by increasing the humidity, the conductivity increases, resulting in a higher dielectric constant [145]. A ceramic thick film humidity sensor of MnZn ferrite based on interdigitated electrodes has been reported by Arshaka et al. [162].

The humidity sensing efficiency of a material depends on its microstructural characteristics that are related to the synthesis method [161]. In this sense, the higher sensitivity of $ZnFe_2O_4$ can be attributed to the small grain size, high surface area accessible for adsorption of water vapor and is correlated with the low barrier height. Nanosized $CoFe_2O_4$, $CuFe_2O_4$ and $NiFe_2O_4$ have also been demonstrated to be good sensors for detecting oxidizing gases such as chlorine [60]. Owing to its high H_C values and stability, the sensors and actuators made by $CoFe_2O_4$ are more durable and of extensive use [7]. Magnetostrictive $CoFe_2O_4$ composites attracted interest for the development of magnetoelastic sensors based on high magnetostriction, response to applied stress, chemical inertness and low cost [163]. Furthermore, due to the bifunctional nature in terms of stress sensing and actuation or contraction under the influence of a magnetic field, $CoFe_2O_4$ has been classified as a smart material with various technological applications among position sensors [120].

Fiber optic biosensors offer great advantages over conventional sensors. Due to its technological advantages (high stability, possibility to be easily removed from the reaction medium, and reuse), the NPs immobilized in glucose oxidase are of great significance for the development of fiber optic glucose oxidase sensors, widely used in food technology, fermentation products, and glucose biosensor [164]. Besides, the immobilization of glucose oxidase sensor on functionalized $CoFe_2O_4@SiO_2$ NPs via cross-linking with glutaraldehyde can be an effective way to produce the ideal immobilized enzyme [164].

A carbon paste electrode containing $ZnFe_2O_4$ NPs was proven to have high sensitivity and good reproducibility in detecting trace levels of 5-fluorourcile in drug samples [165].

3.9. Dielectric Applications

Generally, the dielectric structure involves well conducting grains, separated by low conductivity grain boundaries [166,167]. The dielectric properties of spinel ferrites are determined by structural homogeneity, cation distribution, particle size, density, and porosity [167]. These properties are highly dependent on the synthesis method and thermal treatment parameters, such as temperature, duration, or heating and cooling rate [74].

In case of $CoFe_2O_4$, the small Co^{2+} ion occupying the tetrahedral site, reduces the lattice constant, facilitates the electron hopping between Fe^{3+} and Fe^{2+} ions and acts as a source of charge carrier enhancing the dielectric behavior of $CoFe_2O_4$ [8]. At low frequencies, the polarization in $CoFe_2O_4$ is determined by the hopping of charge carriers, that accumulates and produces polarization upon reaching the grain boundaries, leading to high dielectric constants. In contrast, at the higher frequencies range, the hopping of charge carriers is not able to follow the alternating current induced field, and are thus incompletely polarized, resulting low dielectric constants [166,167]. The dielectric constant decreases by increasing the grain size, which further decreases the grain boundary between the small grains [118]. The dielectric constant is high at lower frequencies, and decreases with increase of frequency [167]. This decrease becomes constant beyond a certain frequency as only the dielectric polarization contributes to dielectric constant [25]. The dielectric constant gradually also increases with increasing temperature due to dielectric polarization [168]. At low temperature, the charge carriers are unable to orient along the applied electric field direction, therefore the polarization is weak and the dielectric constant is very low [168]. However, as temperature increases, a high number of charge carriers liberate and contribute to polarization, that further increases the dielectric constant [168]. The charge carriers exchange between different valence states of elements present in nanosized ferrites highly influences the polarization [166]. High dielectric constants at high frequencies are attributed to the occurrence of space charge polarization due to inhomogeneous dielectric structure of grain size and impurities [169]. Gopalan et al. [170] reported lower dielectric constant values of $CoFe_2O_4$ NPs prepared by the sol-gel method compared to bulk $CoFe_2O_4$. The $CoFe_2O_4@SiO_2$ NCs with controlled magnetic and dielectric properties are promising candidates for biological and high frequency applications [30]. $CuFe_2O_4$ is a potential candidate for microwave devices due to the adequate magnetic and dielectric properties at high frequency [27].

$NiFe_2O_4$ has also a dielectric structure with grains and grain boundaries of different conducting properties [171]. The exchange electron between Fe^{2+} and Fe^{3+} ions and the hole that transfer between Ni^{3+} and Ni^{2+} ions assure the electrical conduction and dielectric polarization [171]. By increasing the frequencies, the electron/hole exchange frequency will not be able to follow the applied electric field, resulting in lower polarization [171].

3.10. Antimicrobial Applications

Iron plays a key role in microbial pathogenesis, many organisms using Fe sequestration as a defense against infection [172]. The Fe(II)- and Fe(III)-resistant microorganisms showed unexpected resistance to a range of antibiotics such as ampicillin, chloramphenicol, rifampicin, sulfanilamide and tetracycline. Ferrite NPs exhibit good antibacterial properties against *Bacillus cereus, Escherichia coli, Staphylococcus aureus, Pseudomonas aeruginosa, Serratia marcescens*, and *Candida albicans*. The strain of *Escherichia coli* was capable of evolving resistance to excess Fe mainly through changes in the uptake of Fe [172].

$CoFe_2O_4$ NPs are well-known candidates for biomedical applications due to the antimicrobial activity against many human pathogens [55]. The high surface-to-volume ratios and nanoscale size of $CoFe_2O_4$ NPs improves their reaction with the pathogenic microbes [173]. $CoFe_2O_4@SiO_2@Ag$ nanocomposite and its combination with Streptomycin exhibited significant antibacterial activity against both Gram-positive and Gram-negative bacteria. The deposition of Ag NPs on $CoFe_2O_4@SiO_2$ can prevent agglomeration. Moreover, it can be easily recovered from solution after disinfection, by applying of an external magnetic field due to the presence of magnetic core in the composite [174].

$CoFe_2O_4$ adheres to the membranes of microorganisms, thus increasing the lag stage of the bacterial growth period, spreading the production time of microorganisms and increasing the bacterial cell division. The antimicrobial activity of $CoFe_2O_4$ is due to its produced effective oxides (i.e., superoxide (O^{2-}) and hydrogen peroxide (H_2O_2)) [173]. The ferrite NPs antibacterial effect on unicellular fungi is dependent on the diffusion flow of an effective oxide within the bacteria cell surface [173].

The antimicrobial activity of $CoFe_2O_4$ NPs was studied against multidrug resistant clinical pathogens (*Staphylococcus aureus, Escherichia coli, Candida parapsilosis* and *Candida albicans*) by assessing the colony forming units [175,176]. Furthermore, $CoFe_2O_4$ NPs functionalized with oleine ($CoFe_2O_4@Ole$) and lysine ($CoFe_2O_4@Lys$) demonstrated high efficiency against all tested microorganisms most probably due to (i) $CoFe_2O_4@Lys$ NPs can easily interact electrostatically with negatively charged *S. aureus* and *E. coli* cell walls, causing destruction of the cytoplasm; (ii) positively charged $CoFe_2O_4@Lys$ NPs and negatively charged $CoFe_2O_4@Ole$ NPs by can affect bacteria cell homoeostasis. The high antimicrobial efficiency of $CoFe_2O_4$@Ole NPs could be due to their higher stability compared to $CoFe_2O_4$@Lys NPs or to a negative curvature wrapping of anionic membranes [175].

The $ZnFe_2O_4$ nanomaterials also display good antibacterial activity against *Lactobacillus*, *Bacillus cereus* (Gram-positive), *Escherichia coli*, *Aeromonas hydrophila* and *Vibrio harveyi* (Gram-negative) bacterial strains [145,146,155]. The low crystallite size and surface area play a critical role in the antimicrobial activity against tested pathogenic microbes. The $ZnFe_2O_4$ NPs had no activity against *Escherichia coli* and was only active against *Staphylococcus aureus* and *Pseudomonas aeruginosa* [173].

$NiFe_2O_4$ had no activity against *E. coli*, but was active against *S. aureus* and *P. aeruginosa* [176]. In addition, the antibacterial effect of $NiFe_2O_4$@carbon nanocomposite on the degradation of *P. aeruginosa* bacteria was rapid and sensitive [95].

3.11. Biomedical Applications

The biomedical applications require non-agglomerated and stable aqueous dispersion of magnetic NPs with high M_S values and good biocompatibility [177]. The magnetic NPs can be highly effective for magnetic fluid hyperthermia, drug release and thermal excitation of metabolic pathways within a single cell [178]. In this regard, the $MnFe_2O_4$ NPs have attracted considerable attention in biomedicine due their easy synthesis process, controllable size, high magnetization value, superparamagnetic nature, ability to be monitored by external magnetic field, surface manipulation capability, and greater biocompatibility [3,179]. Moreover, the surface modification of $MnFe_2O_4$ NPs with mesoporous SiO_2 or its incorporation into mesoporous SiO_2 nanosphere could play an important role in the stabilization of NPs in water, in the enhancement of its biocompatibility, and in reducing the agglomeration and degradation of $MnFe_2O_4$, respectively [3].

The physical characteristics of $CoFe_2O_4$, especially high stability, led to extensive attention regarding potential biomedical applications [173]. Generally, $CoFe_2O_4$ has been used for drug delivery, imaging factor and therapy of brain tumors [173]. At high concentrations of $CoFe_2O_4$ NPs, metal ions are released and some of them enter the cells leading to cytotoxicity, according to the Trojan horse mechanism. A possible explanation could be the prevention of cell transcription and protein synthesis and subsequent altered cellular function [31]. The biocompatible SiO_2 coating could remove or reduce the adverse effects. By coating $CoFe_2O_4$ NPs, their adhesion decreased due to the negative charge of SiO_2 matrix, and they displayed superparamagnetic behavior with a low M_S value and higher compatibility, which made them suitable for various medical applications (drug delivery, specifically cancer cells, magnetic resonance imaging contrast agent for cancer diagnosis) [31,180].

Hyperthermia is a noninvasive treatment procedure for oncological pathologies in which both healthy and carcinoma cells of a living tissue are exposed to necrosis, by prolonged overheating (>43 °C) [181]. The existing tumor heating methods (hot water, microwave, infrared ray, ultraviolet ray, etc.) are ineffective for deep-seated cancers, but the magnetic induction hyperthermia, i.e., via magnetic NPs, may be effective in these types of tumor due to the selective heating and destruction of tumor tissue, with minimum collateral damage [178,182]. Magnetic hyperthermia is an emerging adjuvant therapy for malign tumors, where the appropriate magnetic NPs are located under a magnetic field and the heat released may remove the cancerous tissue, at an optimum temperature of 41–46 °C [15,123]. Ferrimagnetic materials own hysteretic properties under time-varying magnetic field, which lead to magnetically induced heating [183]. Moreover, the magnetic NPs have low toxicity, are biocompatible and well-tolerated by the living organisms, leading to a simple and fast analysis of specific absorption rate [177]. Some studies reported an effective technique of inserting ferrimagnetic NPs in a tumor region based on the magnetophoresis, a phenomenon caused by a magnetic field gradient on magnetically induced magnetic moment of particle [180,183]. Moreover, magnetophoresis is used in numerous commercial and industrial processes for the separation of magnetic NPs dangling in fluids [183].

However, the main challenge of hyperthermia is to lower the damage to nearby normal tissues. In this regard, NPs must be concentrated at the tumor site rather than in the tumor surroundings [50]. The uncoated magnetic $NiFe_2O_4$ NPs have the potential to be used in cell differentiable hyperthermia agents due to their high value of cell survival rate (85%) and non-cytotoxicity under different pH levels (pH = 7 normal cell and pH = 6 tumor cell) [15]. Further, $ZnFe_2O_4$ is a good candidate for hyperthermia due to its low toxicity [15,123]. The nanosized ferrite spinel's enzyme-like activities include peroxidase, oxidase, and catalase, and their applications as enzyme mimetics in biosensing, molecular detection, cancer therapy, and drug delivery were recently reviewed [184].

The hyperthermia measurements $CoFe_2O_4$ and $MnFe_2O_4$ NPs were performed under various conditions, in order to normalize the obtained results by removing their dependence on magnetic field frequency and amplitude. $MnFe_2O_4$ NPs obtained in gelatin medium at 175 °C display the best reported heating efficiency, with values comparable to those of commercial magnetic NPs [185]. $CoFe_2O_4$ NPs with narrow size distribution and small particle size (~10 nm), and hence small hydrodynamic diameter, are successfully used to produce high magnetic hyperthermia within short duration [181]. Moreover, the magnetic $CoFe_2O_4@MnFe_2O_4$ NPs have proven to be superior to the individual NPs for producing effective heating in hyperthermia [178].

$ZnFe_2O_4$ type bioactive glass ceramics possess both the magnetic properties that generate adequate heat and the ability to bond to natural tissues (via hydroxyapatite layer) [182]. Consequently, this type of materials can be used for the hyperthermia treatment of cancer, but also as a substitute for a cancerous/ damaged bone. Such magnetic heat generation materials depend on various factors such as the structure and magnetic characteristics of the material, strength and frequency of alternating magnetic field, quantity of implant, etc. [182].

The synthesis of $CoFe_2O_4@DMSA/DOX$ NPs to improve the efficiency of magnetic $CoFe_2O_4$ in therapeutic applications, by functionalizing its surface with meso-2,3-dimercaptosuccinic acid, followed by conjugation with the anticancer agent, doxorubucin was also reported [186]. As expected, the experimental results demonstrate that the combined thermal and chemotherapy effect induced by $CoFe_2O_4@DMSA/DOX$ NPs displays an excellent antitumor efficacy by mitochondrial membrane disruption [186]. The well dispersed spherical magnetic $ZnFe_2O_4$ NPs exhibit a better heat efficiency compared to the aggregated polyhedral magnetic $ZnFe_2O_4$ NPs, being able to produce a threshold hyperthermia temperature of 42–45 °C in a short time, a key feature for magnetic hyperthermia applications [187].

The cytotoxicity of NPs considered as potential drug delivery systems must be taken into account, especially since it may vary with the type of cell line, exposure dosage, exposure time, particle size and aggregation tendency [188]. The in vitro estimation of cytotoxicity using different cancer cell lines (HeLa, cervical cancer cell and PC-3, prostate cancer cell) revealed good biocompatibility of $NiFe_2O_4$ NPs synthesized by co-precipitation and subsequent thermal annealing and less toxic effect to normal cell L929 [188].

The superparamagnetic response and biocompatibility of magnetic NPs make them potential candidates as contrast agents in magnetic resonance imaging (MRI) and tracer agents in magnetic particle imaging (MPI). Due to its soft magnetic behavior, $NiFe_2O_4$ is used as a contrast agent in MRI [189]. $NiFe_2O_4@PAA$ composites (PAA - polyacrylic acid) could be considered as potential tracer agents for MPI, outperforming the commercial, commonly used tracer agents due to the minimum relaxation time of 3.10 µs and high resolution of 7.75 mT, [189]. Besides its use in photothermal and sonodynamic therapy of melanoma, $MnFe_2O_4/C$ nanocomposite may also be used as a novel theranostic agent in nanomedicine [190]. In this regard, the intratumoral injection of $MnFe_2O_4/C$ may induce deep tumor tissue necrosis, while the imaging studies endorse $MnFe_2O_4/C$ as a contrast agent for MRI with a dose-dependent decrease in the signal intensity [189,190]. However, comprehensive studies on the in vitro and in vivo biodegradability and pharmacokinetics of $MnFe_2O_4/C$ nanocomposite are necessary for additional investigations and evaluation [189,190].

4. Conclusions

Nanosized $CoFe_2O_4$, $MnFe_2O_4$, $ZnFe_2O_4$, $NiFe_2O_4$, and $CuFe_2O_4$ NPs have received impressive attention in the last decade due to their special properties, such as chemical, mechanical, and thermal stability, large coercivity, high anisotropy constant and Curie temperature, moderate saturation magnetization, high electrical resistance, and low eddy current loss. Amongst all the reviewed synthesis methods, the sol-gel and chemical co-precipitation techniques stand as superlative routes for synthesizing fine, homogenous, nanostructured ferrites. However, several unconventional methods allow the cost-efficient preparation of high-quality NPs. Although bulk ferrites remain a key group of magnetic materials, due to their advantageous properties, the investigated nanostructured ferrites are used in many areas, including material sciences, engineering, physics, chemistry, biology, and medicine. Notably, these magnetic ferrite NPs have displayed outstanding results in the field of medicine because of their low toxicity, biocompatibility, and ability to be easily manipulated using a magnetic field.

Author Contributions: T.D., E.A.L. and O.C. conceived and designed the work; T.D., E.A.L. and O.C. wrote the manuscript. All authors have read and agreed to the published version of the manuscript.

Funding: This work was supported by a grant of the Romanian National Authority for Scientific Research CNCS-UEFISCDI, project number PN-III-P2-2.1-PED-2019-3664.

Data Availability Statement: The data presented in this study are available on request from the corresponding author.

Conflicts of Interest: The authors declare no conflict of interest. The funders had no role in the design of the study; in the collection, analyses, or interpretation of data; in the writing of the manuscript, or in the decision to publish the results.

References

1. Chand, P.; Vaish, S.; Kumar, P. Structural, optical and dielectric properties of transition metal (MFe_2O_4; M = Co, Ni and Zn) nanoferrites. *Phys. B Condens Matter.* **2017**, *524*, 53–63. [CrossRef]
2. Jeevanandam, J.; Barhoum, A.; Chan, Y.S.; Dufresne, A.; Danquah, M.K. Review on nanoparticles and nanostructured materials: History, sources, toxicity and regulations. *Beilstein. J. Nanotechnol.* **2018**, *9*, 1050–1074. [CrossRef]
3. Asghar, K.; Qasim, M.; Das, D. Preparation and characterization of mesoporous magnetic $MnFe_2O_4@mSiO_2$ nanocomposite for drug delivery application. *Mater. Today Proc.* **2020**, *26*, 87–93. [CrossRef]
4. Sivakumar, P.; Ramesh, R.; Ramanand, A.; Ponnusamy, S.; Muthamizhchelvan, C. Synthesis and characterization of $NiFe_2O_4$ nanoparticles and nanorods. *J. Alloy Comp.* **2013**, *563*, 6–11. [CrossRef]
5. Ozçelik, B.; Ozçelik, S.; Amaveda, H.; Santos, H.; Borrell, C.J.; Saez-Puche, R.; de la Fuente, G.F.; Angurel, L.A. High speed processing of $NiFe_2O_4$ spinel using laser furnance. *J. Materiomics* **2020**, *6*, 661–670. [CrossRef]
6. Džunuzović, A.S.; Ilić, N.I.; Vijatović Petrović, M.M.V.; Bobić, J.D.; Stojadinović, B.; Dohčević-Mitrović, Z.; Stojanović, B.D. Structure and properties of Ni–Zn ferrite obtained by auto-combustion method. *J. Magn. Magn. Mater.* **2018**, *374*, 245–251. [CrossRef]
7. Kaur, H.; Singh, A.; Kumar, A.; Ahlawat, D.S. Structural, thermal and magnetic investigations of cobalt ferrite doped with Zn^{2+} and Cd^{2+} synthesized by auto combustion method. *J. Magn. Magn. Mater.* **2019**, *474*, 505–511. [CrossRef]
8. Naik, A.B.; Naik, P.P.; Hasolkar, S.S.; Naik, D. Structural, magnetic and electrical properties along with antifungal activity & adsorption ability of cobalt doped manganese ferrite nanoparticles synthesized using combustion route. *Ceram. Int.* **2020**, *46*, 21046–21055.
9. Xiong, Q.Q.; Tu, J.P.; Shi, S.J.; Liu, X.Y.; Wang, X.L.; Gu, C.D. Ascorbic acid-assisted synthesis of cobalt ferrite ($CoFe_2O_4$) hierarchical flower-like microspheres with enhanced lithium storage properties. *J. Power Sources* **2014**, *256*, 153–159. [CrossRef]
10. Li, X.; Sun, Y.; Zong, Y.; Wei, Y.; Liu, X.; Li, X.; Peng, Y.; Zheng, X. Size-effect induced cation redistribution on the magnetic properties of well-dispersed $CoFe_2O_4$ nanocrystals. *J. Alloy Comp.* **2020**, *841*, 155710. [CrossRef]
11. Torkian, S.; Ghasemi, A.; Razavi, R.S. Cation distribution and magnetic analysis of wideband microwave absorptive $Co_xNi_{1-x}Fe_2O_4$ ferrites. *Ceram. Int.* **2017**, *43*, 6987–6995. [CrossRef]
12. Mahala, C.; Sharma, M.D.; Basu, M. 2D nanostructures of $CoFe_2O_4$ and $NiFe_2O_4$: Efficient oxygen evolution catalyst. *Electrochim. Acta* **2018**, *273*, 462–473. [CrossRef]

13. Manikandan, A.; Sridhar, R.; Arul, S.A.; Ramakrishna, S. A simple aloe vera plant-extracted microwave and conventional combustion synthesis: Morphological, optical, magnetic and catalytic properties of $CoFe_2O_4$ nanostructures. *J. Molec. Struct.* **2014**, *1076*, 188–200. [CrossRef]

14. Salunkhe, A.B.; Khot, V.M.; Phadatare, M.R.; Thorat, N.D.; Joshi, R.S.; Yadav, H.M.; Pawar, S.H. Low temperature combustion synthesis and magnetostructural properties of Co-Mn nanoferrites. *J. Magn. Magn. Mater.* **2014**, *352*, 91–98. [CrossRef]

15. Ghayour, H.; Abdellahi, M.; Ozada, N.; Jabbrzare, S.; Khandan, A. Hyperthermia application of zinc doped nickel ferrite nanoparticles. *J. Phys Chem. Solids* **2017**, *111*, 464–472. [CrossRef]

16. Kremenović, A.; Antić, B.; Vulić, P.; Blanuša, J.; Tomic, A. $ZnFe_2O_4$ antiferromagnetic structure redetermination. *J. Magn. Magn. Mater.* **2017**, *426*, 264–266. [CrossRef]

17. Manikandan, A.; Durka, M.; Arul, S.A. Magnetically recyclable spinel $Mn_xZn_{1-x}Fe_2O_4$ $(0.0 \leq x \geq 0.5)$. *Adv. Sci. Eng. Med.* **2015**, *7*, 33–46. [CrossRef]

18. Ge, Y.-C.; Wang, Z.-L.; Yi, M.-Z.; Ran, L.-P. Fabrication and magnetic transformation from paramagnetic to ferrimagnetic of $ZnFe_2O_4$ hollow spheres. *Trans. Nonferrous Met. Soc. China* **2019**, *29*, 1503–1509.

19. Feng, S.; Yang, W.; Wang, Z. Synthesis of porous $NiFe_2O_4$ microparticles and its catalytic properties for methane combustion. *Mater. Sci. Eng. B* **2011**, *176*, 1509–1512. [CrossRef]

20. Alarifi, A.; Deraz, N.M.; Shaban, S. Structural, morphological and magnetic properties of $NiFe_2O_4$ nano-particles. *J. Alloy Comp.* **2009**, *486*, 501–506. [CrossRef]

21. Shanmugavel, T.; Raj, S.G.; Rajarajan, G.; Kumar, G.R. Tailoring the structural and magnetic properties and of nickel ferrite by auto combustion method. *Proc. Mat. Sci.* **2014**, *6*, 1725–1730. [CrossRef]

22. Shetty, K.; Renuka, L.; Nagaswarupa, H.P.; Nagabhushana, H.; Anantharaju, K.S.; Rangappa, D.; Prashantha, S.C.; Ashwini, K. A comparative study on $CuFe_2O_4$, $ZnFe_2O_4$ and $NiFe_2O_4$: Morphology, impedance and photocatalytic studies. *Mater. Today Proc.* **2017**, *4*, 11806–11815. [CrossRef]

23. Iqbal, M.J.; Yaqub, N.; Sepiol, B.; Ismail, B. A study of the influence of crystallite size on the electrical and magnetic properties of $CuFe_2O_4$. *Mat. Res. Bull.* **2011**, *46*, 1837–1842. [CrossRef]

24. Shilpa Amulya, M.A.; Nagaswarupta, H.P.; Anil Kumar, M.R.; Ravikumar, C.R.; Kusuma, K.B.; Prashantha, S.C. Evaluation of bifunctional applications of $CuFe_2O_4$ nanoparticles synthesized by a sonochemical method. *J. Phys. Chem. Solids* **2021**, *148*, 109756. [CrossRef]

25. Dar, M.A.; Varshney, D. Effect of d-block element Co^{2+} substitution on structural, Mössbauer and dielectric properties of spinel copper ferrite. *J. Magn. Magn. Mater.* **2017**, *436*, 101–112. [CrossRef]

26. Mohanty, D.; Mallick, P.; Biswall, S.K.; Behera, B.; Mohapatra, R.K.; Behera, A.; Satpathy, S.K. Investigation of structural, dielectric and electrical properties of $ZnFe_2O_4$. *Mater Today Proc.* **2020**, *33*, 4971–4975. [CrossRef]

27. Mohanty, D.; Satpathy, S.K.; Behera, B.; Mohapatra, R.K. Dielectric and frequency dependent transport properties in magnesium doped $CuFe_2O_4$ composite. *Mater Today Proc.* **2020**, *33*, 5226–5231. [CrossRef]

28. Sivakumar, A.; Dhas, S.S.J.; Dhas, S.A.M.B. Assessment of crystallographic and magnetic phase stabilities on $MnFe_2O_4$ nano crystalline materials at shocked conditions. *Solid State Sci.* **2020**, *107*, 106340. [CrossRef]

29. Junlabhut, P.; Nuthongkum, P.; Pechrapa, W. Influences of calcination temperature on structural properties of $MnFe_2O_4$ nanopowders synthesized by co-precipitation method for reusable absorbent materials. *Mater Today Proc.* **2018**, *5*, 13857–13864. [CrossRef]

30. Nadeem, K.; Zev, F.; Azeem Abid, M.; Mumtaz, M.; Anis ur Rehman, M. Effect of amorphous silica matrix on structural, magnetic, and dielectric properties of cobalt ferrite/silica nanocomposites. *J. Non Cryst. Solids* **2014**, 30–45. [CrossRef]

31. Gharibshahian, M.; Mirzaee, O.; Nourbakhsh, M.S. Evaluation of superparamagnetic and biocompatible properties of mesoporous silica coated cobalt ferrite nanoparticles synthesized via microwave modified Pechini method. *J. Magn. Magn. Mater.* **2017**, *425*, 48–56. [CrossRef]

32. Zate, M.K.; Raut, S.D.; Shirsat, S.D.; Sangale, S.; Kadam, A.S. Ferrite nanostructures: Synthesis methods. In *Spinel Ferrite Nanostructures for Energy Storage Devices*, 1st ed.; Mane, R., Jadhav, V., Eds.; Elsevier: Amstardam, The Netherlands, 2020. [CrossRef]

33. Vedrtnam, A.; Kalauni, K.; Dubey, S.; Kumar, A. A comprehensive study on structure, properties, synthesis and characterization of ferrites. *AIMS Mat. Sci.* **2020**, *7*, 800–835.

34. Vinosha, P.A.; Manikandan, A.; Preetha, A.C.; Dinesh, A.; Slimani, Y.; Almessiere, M.A.; Baykal, A.; Xavier, B.; Nirmala, F.G. Review on recent advances of synthesis, magnetic properties, and water treatment applications of cobalt ferrite nanoparticles and nanocomposites. *J. Supercond. Nov. Magn.* **2021**, *34*, 995–1018. [CrossRef]

35. Masunga, N.; Mmelesi, O.K.; Kefeni, K.K.; Mamba, B.B. Recent advances in copper ferrite nanoparticles and nanocomposites synthesis, magnetic properties and application in water treatment: Review. *J. Environ. Chem. Eng.* **2019**, *7*, 103179. [CrossRef]

36. Kumar, M.; Dosanjh, S.J.; Singh, J.; Monir, K.; Singh, H. Review on magnetic nano ferrites and their composites as an alternative in waste water treatment: Synthesis, modifications and applications. *Environ. Sci. Water Res. Technol.* **2020**, *6*, 491–514. [CrossRef]

37. Kefeni, K.K.; Mamba, B.B. Photocatalytic application of spinel ferrite nanoparticles and nanocomposites in wastewater treatment: Review. *Sustain. Mater. Tech.* **2020**, *23*, e00140. [CrossRef]

38. Kharisov, B.I.; Rasika Dias, H.V.; Kharissova, O.V. Mini-review: Ferrite nanoparticles in the catalysis. *Arab. J. Chem.* **2019**, *12*, 1234–1246. [CrossRef]

39. Dalawai, S.P.; Kumar, S.; Al Saad Aly, M.; Khan, M.Z.H.; Xing, R.; Vasambekar, P.N.; Liu, S. A review of spinel-type of ferrite thick flm technology: Fabrication and application. *J. Mat. Sci: Mat. Electr.* **2019**, *30*, 7752–7779.

40. Kefeni, K.K.; Msagati, T.A.M.; Mamba, B.B. Ferrite nanoparticles: Synthesis, characterisation and applications in electronic device. *Mat. Sci. Eng. B* **2017**, *215*, 37–55. [CrossRef]

41. Dippong, T.; Cadar, O.; Levei, E.A.; Deac, I.G. Microstructure, porosity and magnetic properties of $Zn_{0.5}Co_{0.5}Fe_2O_4/SiO_2$ nanocomposites prepared by sol-gel method using different polyols. *J. Magn. Magn. Mater.* **2020**, *498*, 166168. [CrossRef]

42. Dippong, T.; Levei, E.A.; Deac, I.G.; Neag, E.; Cadar, O. Influence of Cu^{2+}, Ni^{2+}, and Zn^{2+} ions doping on the structure, morphology, and magnetic properties of Co-ferrite embedded in SiO_2 matrix obtained by an innovative sol-gel route. *Nanomaterials* **2020**, *10*, 580. [CrossRef]

43. Dippong, T.; Levei, E.A.; Deac, I.G.; Goga, F.; Cadar, O. Investigation of structural and magnetic properties of $Ni_xZn_{1-x}Fe_2O_4/SiO_2$ $(0 \leq x \leq 1)$ spinel-based nanocomposites. *J. Anal. Appl. Pyrol.* **2019**, *144*, 104713. [CrossRef]

44. Dippong, T.; Deac, I.G.; Cadar, O.; Levei, E.A.; Petean, I. Impact of Cu^{2+} substitution by Co^{2+} on the structural and magnetic properties of $CuFe_2O_4$ synthesized by sol-gel route. *Mater. Caract.* **2020**, *163*, 110248. [CrossRef]

45. Dippong, T.; Cadar, O.; Deac, I.G.; Lazar, M.; Borodi, G.; Levei, E.A. Influence of ferrite to silica ratio and thermal treatment on porosity, surface, microstructure and magnetic properties of $Zn_{0.5}Ni_{0.5}Fe_2O_4/SiO_2$ nanocomposites. *J. Alloy. Comp.* **2020**, *828*, 15409. [CrossRef]

46. Dippong, T.; Levei, E.A.; Cadar, O. Formation, structure and magnetic properties of $MFe_2O_4@SiO_2$ (M = Co, Mn, Zn, Ni, Cu) nanocomposites. *Materials* **2021**, *14*, 1139. [CrossRef] [PubMed]

47. Cadar, O.; Dippong, T.; Senila, M.; Levei, E.A. Progress, challenges and opportunities in divalent transition metal doped cobalt ferrites nanoparticles applications. *IntechOpen* **2020**. [CrossRef]

48. Hashim, M.; Alimuddin; Kumar, S.; Koo, B.H.; Shirsath, S.E.; Mohammed, E.M.; Shah, J.; Kotnala, R.K.; Choi, H.K.; Chung, H.; et al. Structural, electrical and magnetic properties of Co–Cu ferrite nanoparticles. *J. Alloy. Comp.* **2012**, *518*, 11–18. [CrossRef]

49. Rao, K.R.; Nayakulu, S.V.R.; Varma, M.C.; Choudary, G.S.V.R.K.; Rao, K.H. Controlled phase evolution and the occurrence of single domain $CoFe_2O_4$ nanoparticles synthesized by PVA assisted sol-gel method. *J. Magn. Magn. Mater.* **2018**, *451*, 602–608.

50. Ajinka, N.; Yu, X.; Kaithal, P.; Luo, H.; Somani, P.; Ramakrishna, S. Magnetic iron oxide nanoparticle (IONP) synthesis to applications: Present and Future. *Materials* **2020**, *13*, 4644. [CrossRef]

51. Amirabadizadeh, A.; Salighe, Z.; Sarhaddi, R.; Lotfollahi, Z. Synthesis of ferrofluids based on cobalt ferrite nanoparticles: Influence of reaction time on structural, morphological and magnetic properties. *J. Magn. Magn. Mater.* **2017**, *434*, 78–85. [CrossRef]

52. Yousefi, M.H.; Manouchehri, S.; Arab, A.; Mozaffari, M.; Amiri, G.R.; Amighian, J. Preparation of cobalt–zinc ferrite $(Co_{0.8}Zn_{0.2}Fe_2O_4)$ nanopowder via combustion method and investigation of its magnetic properties. *Mat. Res. Bull.* **2010**, *45*, 1792–1795. [CrossRef]

53. Hema, E.; Manikandan, A.; Gayathri, M.; Durka, M.; Arul Antony, S.; Venkatraman, B.R. Role of Mn^{2+}-doping on structural, morphological, optical, magnetic and catalytic properties of spinel $ZnFe_2O_4$ nanoparticles. *J. Nanosci. Nanotechnol.* **2016**, *16*, 5929–5943. [CrossRef] [PubMed]

54. Baykal, A.; Kasapoğlu, N.; Köseoğlu, Y.; Başaran, A.C.; Kavas, H.; Toprak, M.S. Microwave-induced combustion synthesis and characterization of $Ni_xCo_{1-x}Fe_2O_4$ Nanocrystals (x = 0.0, 0.4, 0.6, 0.8, 1.0). *Cent. Eur. J. Chem.* **2008**, *6*, 125–130. [CrossRef]

55. Naik, M.M.; Naik, H.S.B.; Nagaraju, G.; Vinuth, M.; Vinu, K.; Viswanath, R. Green synthesis of zinc doped cobalt ferrite nanoparticles: Structural, optical, photocatalytic and antibacterial studies. *Nano Struct. Nano Obj.* **2019**, *19*, 100322. [CrossRef]

56. Bakhshi, H.; Vahdati, N.; Sedghi, A.; Mozharivskyj, Y. Comparison of the effect of nickel and cobalt cations addition on the structural and magnetic properties of manganese-zinc ferrite nanoparticles. *J. Magn. Magn. Mater.* **2019**, *474*, 56–62. [CrossRef]

57. Houshiar, M.; Zebhi, F.; Razi, Z.J.; Alidoust, A.; Askari, Z. Synthesis of cobalt ferrite ($CoFe_2O_4$) nanoparticles using combustion, coprecipitation, and precipitation methods: A comparison study of size, structural and magnetic properties. *J. Magn. Magn. Mater.* **2014**, *371*, 43–48. [CrossRef]

58. Vinosha, P.A.; Minikandan, A.; Ceicilia, A.S.J.; Dinesh, A.; Nirmala, G.F.; Preetha, A.C.; Slimani, Y.; Almessiere, M.A.; Baycal, A.; Xavier, B. Review on recent advances of zinc substituted cobalt ferrite nanoparticles: Synthesis characterization and diverse applications. *Ceram. Int.* **2021**, *47*, 10512–10535. [CrossRef]

59. Thakur, P.; Chahar, D.; Taneja, S.; Bhalla, N.; Thakur, A. A review on MnZn ferrites: Synthesis, characterization and applications. *Ceram. Int.* **2020**, *46*, 15740–15763. [CrossRef]

60. Jeseentharani, V.; George, M.; Jeyaraj, B.; Dayalan, A.; Nagaraj, K.S. Synthesis of metal ferrite (MFe_2O_4, M = Co, Cu, Mg, Ni, Zn) nanoparticles as humidity sensor materials. *J. Exp. Nanosci.* **2013**, *8*, 358–370. [CrossRef]

61. Qin, R.; Li, F.; Liu, L.; Jiang, W. Synthesis of well-dispersed $CoFe_2O_4$ nanoparticles via PVA-assisted low-temperature solid state process. *J. Alloys Comp.* **2009**, *482*, 508–511. [CrossRef]

62. Varma, P.C.R.; Manna, R.S.; Benerjee, D.; Varma, M.R.; Suresh, K.G.; Nigam, A.K. Magnetic properties of $CoFe_2O_4$ synthesized by solid state, citrate precursor and polymerized complex methods: A comparative study. *J. Alloys Comp.* **2008**, *453*, 298–303. [CrossRef]

63. Lemine, O.M.; Bououdina, M.; Sajieddine, M.; Al-Saie, A.M.; Shafi, M.; Khatab, A.; Al-Hilali, M.; Henini, M. Synthesis, structural, magnetic and optical properties of nanocrystalline $ZnFe_2O_4$. *Phys. B Condens Matter.* **2011**, *406*, 1989–1994. [CrossRef]

64. Atif, M.; Turtelli, R.S.; Grössinger, R.; Siddique, M.; Nadeem, M. Effect of Mn substitution on the cation distribution and temperature dependence of magnetic anisotropy constant in $Co_{1-x}Mn_xFe_2O_4$ ($0.0 \leq x \leq 0.4$) ferrites. *Ceram. Int.* **2014**, *40*, 471–748. [CrossRef]

65. Aslibeiki, B.; Kameli, P.; Salamati, H.; Eshraghi, M.; Tahmasebi, T. Superspin glass state in $MnFe_2O_4$ nanoparticles. *J. Magn. Magn. Mater.* **2010**, *322*, 2929–2934. [CrossRef]

66. Chia, C.H.; Zakaria, S.; Yusoff, M.; Goh, S.C.; Hawa, C.Y.; Ahmadi, S.; Huang, N.M.; Lim, H.M. Size and crystallinity-dependent magnetic properties of $CoFe_2O_4$ nanocrystals. *Ceram. Int.* **2010**, *36*, 605–609. [CrossRef]

67. Zahraei, M.; Monshi, A.; del Puerto Morales, M.; Shahbazi-Gahrouei, D.; Amirnasr, M.; Behdadfar, M. Hydrothermal synthesis of fine stabilized superparamagnetic nanoparticles of Zn^{2+} substituted manganese ferrite. *J. Magn. Magn. Mater.* **2015**, *393*, 429–436. [CrossRef]

68. Prabhakaran, T.; Mangalaraja, R.V.; Denardin, J.C.; Jimenez, J.A. The effect of reaction temperature on the structural and magnetic properties of nano $CoFe_2O_4$. *Ceram. Int.* **2017**, *43*, 5599–5606. [CrossRef]

69. Senapati, K.K.; Borgohain, C.; Phukan, P. Synthesis of highly stable $CoFe_2O_4$ nanoparticles and their use as magnetically separable catalyst for Knoevenagel reaction in aqueous medium. *J. Mol. Catal. A Chem.* **2011**, *339*, 24–31. [CrossRef]

70. Singh, A.; Pathak, S.; Kumar, P.; Sharma, P.; Rathi, A.; Basheed, G.A.; Maurya, K.K.; Plant, R.P. Tuning the magnetocrystalline anisotropy and spin dynamics in $Co_xZn_{1-x}Fe_2O_4$ ($0 \leq x \leq 1$) nanoferrites. *J. Magn. Magn. Mater.* **2020**, *493*, 165737. [CrossRef]

71. Huixia, F.; Baiyi, C.; Deyi, Z.; Jianqiang, Z.; Lin, T. Preparation and characterization of the cobalt ferrite nano-particles by reverse coprecipitation. *J. Magn. Magn. Mater.* **2014**, *388*, 68–72. [CrossRef]

72. Sajjia, M.; Oubaha, M.; Hasanuzzaman, M.; Olabi, A.G. Developments of cobalt ferrite nanoparticles prepared by the sol–gel process. *Ceram. Int.* **2014**, *40*, 1147–1154. [CrossRef]

73. Wang, L.; Lu, M.; Liu, Y.; Li, J.; Liu, M.; Lin, H. The structure, magnetic properties and cation distribution of $Co_{1-x}Mg_xFe_2O_4/SiO_2$ nanocomposites synthesized by sol–gel method. *Ceram. Int.* **2015**, *41*, 4176–4181. [CrossRef]

74. Zhang, R.; Sun, L.; Wang, Z.; Hao, W.; Cao, E.; Zhang, Y. Dielectric and magnetic properties of $CoFe_2O_4$ prepared by sol-gel auto-combustion method. *Mat. Res. Bull.* **2018**, *98*, 133–138. [CrossRef]

75. Barison, S.; Fabrizio, M.; Fasolin, S.; Montagner, F.; Mortalò, C. A microwave-assisted sol–gel Pechini method for the synthesis of $BaCe_{0.65}Zr_{0.20}Y_{0.15}O_{3-\delta}$ powders. *Mater. Res. Bull.* **2010**, *45*, 1171–1176. [CrossRef]

76. Sharma, S.; Kumar Verma, M.; Sharma, N.D.; Choudhary, N.; Singh, S.; Singh, D. Rare-earth doped Ni–Co ferrites synthesized by Pechini method: Cation distribution and high temperature magnetic studies. *Ceram. Int.* **2021**, *47*, 17510–17519. [CrossRef]

77. Dimesso, L. Pechini Processes: An Alternate Approach of the Sol–Gel Method, Preparation, Properties, and Applications. In *Handbook of Sol-Gel Science and Technology*; Klein, L., Aparicio, M., Jitianu, A., Eds.; Springer International Publishing: Berlin/Heidelberg, Germany, 2016; pp. 1–22.

78. Raut, A.V.; Barkule, R.S.; Shengule, D.R.; Jadhav, K.M. Synthesis, structural investigation and magnetic properties of Zn^{2+} substituted cobalt ferrite nanoparticles prepared by the sol–gel auto-combustion technique. *J. Magn. Magn. Mater.* **2014**, *358*, 87–92. [CrossRef]

79. Mohamed, R.M.; Rashad, M.M.; Haraz, F.A.; Sigmund, W. Structure and magnetic properties of nanocrystalline cobalt ferrite powders synthesized using organic acid precursor method. *J. Magn. Magn. Mater.* **2010**, *322*, 2058–2064. [CrossRef]

80. Barathiraja, C.; Mnikandan, A.; Mohideen Uduman, A.M.; Jayasree, S.; Arul Antony, S. Magnetically recyclable spinel $Mn_xNi_{1-x}Fe_2O_4$ ($x = 0.0$–0.5) nano-photocatalysts: Structural, morphological and opto-magnetic properties. *J. Supercond. Nov. Magn.* **2016**, *29*, 477–486. [CrossRef]

81. Sundararajan, M.; Sailaja, V.; Kennedy, L.J.; Vijaya, J.J. Photocatalytic degradation of rhodamine B under visible light using nanostructured zinc doped cobalt ferrite: Kinetics and mechanism. *Ceram. Int.* **2017**, *43*, 540–548. [CrossRef]

82. Sundararajan, M.; Kennedy, L.J.; Aruldoss, U.; Pasha, S.K.; Vijaya, J.J.; Dunn, S. Microwave combustion synthesis of zinc substituted nanocrystalline spinel cobalt ferrite: Structural and magnetic studies. *Mat. Sci. Semicond Proc.* **2015**, *40*, 1–10. [CrossRef]

83. Singh, R.K.; Rai, B.C.; Prasad, K. Synthesis and Characterization of Copper Substituted Cobalt Ferrite Nanoparticles. *Int. J. Adv. Mater. Sci.* **2012**, *3:2*, 71–76.

84. Garcia Cerda, L.A.; Montemayor, S.M. Synthesis of $CoFe_2O_4$ nanoparticles embedded in a silica matrix by the citrate precursor technique. *J. Mgn. Magn. Mater.* **2005**, *294*, e43–e46. [CrossRef]

85. El-Foulani, A.-H.; Aamouche, A.; Mohseni, F.; Amaral, J.S.; Tobaldi, D.M.; Pullar, R.C. Effect of surfactants on the optical and magnetic properties of cobalt zinc ferrite $Co_{0.5}Zn_{0.5}Fe_2O_4$. *J. Alloys Comp.* **2019**, *774*, 1250–1259. [CrossRef]

86. Naseri, M.G.; Bin Saion, E.; Ahangar, H.A.; Hashim, M.; Shaari, A.H. Synthesis and characterization of manganese ferrite nanoparticles by thermal treatment method. *J. Magn. Magn. Mater.* **2011**, *323*, 1745–1749. [CrossRef]

87. Chakradhary, V.K.; Ansaria, A.; Akhtar., M.J. Design, synthesis, and testing of high coercivity cobalt doped nickel ferrite nanoparticles for magnetic applications. *J. Magn. Magn. Mater.* **2019**, *469*, 674–680. [CrossRef]

88. Soundararajan, D.; Kim, K.H. Synthesis of $CoFe_2O_4$ magnetic nanoparticles by thermal decomposition. *J. Magn.* **2014**, *19*, 5–9. [CrossRef]

89. Peddis, D.; Cannas, C.; Musinu, A.; Ardu, A.; Orru, F.; Fiorani, D.; Laureti, S.; Rinaldi, D.; Muscas, G.; Concas, G.; et al. Beyond the effect of particle size: Influence of $CoFe_2O_4$ nanoparticle arrangements on magnetic properties. *Chem Mater.* **2013**, *25*, 2005–2013. [CrossRef]

90. Kotsikau, D.; Ivanovskaya, M.; Pankov, V.; Fedotova, Y. Structure and magnetic properties of manganeseezinc-ferrites prepared by spray pyrolysis method. *Solid State Sci.* **2015**, *39*, 69–73. [CrossRef]

91. Sarıtaş, S.; Şakar, B.C.; Turgut, E.; Kundakci, M.; Yıldırım, M. Cobalt metal doped magnesium ferrite and zinc ferrite thin films grown by Spray Pyrolysis. *Mater. Today Proc.* **2021**. [CrossRef]

92. Jundale, A.V.; Chorage, G.Y.; Yadav, A.A. Structural, morphological and optical properties of $CoFe_2O_4$ thin film by spray pyrolysis technique. *Mater. Today Proc.* **2021**, *43*, 2678–2681. [CrossRef]

93. Wang, S.; Gao, L. Laser-Driven nanomaterials and laser-enabled nanofabrication for industrial applications. In *Industrial Applications of Nanomaterials*; Sabu, T., Yves, G., Yasir, B.P., Eds.; Elsevier: Amsterdam, The Netherlands, 2019; pp. 181–203. [CrossRef]

94. Bourrioux, S.; Wang, L.P.; Rousseau, Y.; Simon, P.; Habert, A.; Leconte, Y.; Sougrati, T.M.; Stievano, L.; Monconduit, L.; Xu, Z.J.; et al. Evaluation of electrochemical performances of $ZnFe_2O_4/\gamma\text{-}Fe_2O_3$ nanoparticles prepared by laser pyrolysis. *New J. Chem. R. Soc. Chem.* **2017**, *41*, 9236–9243. [CrossRef]

95. Ahmadian-Fard-Fini, S.; Ghanbari, D.; Salavati-Niasari, M. Photoluminescence carbon dot as a sensor for detecting of Pseudomonas aeruginosa bacteria: Hydrothermal synthesis of magnetic hollow $NiFe_2O_4$-carbon dots nanocomposite material. *Composite B* **2019**, *161*, 564–577. [CrossRef]

96. Jalalian, M.; Mirkazemi, S.M.; Alamolhoda, S. The effect of poly vinyl alcohol (PVA) surfactant on phase formation and magnetic properties of hydrothermally synthesized $CoFe_2O_4$ nanoparticles. *J. Magn. Magn. Mater.* **2016**, *419*, 363–367. [CrossRef]

97. Zhao, D.; Wu, X.; Guan, H.; Han, E. Study on supercritical hydrothermal synthesis of $CoFe_2O_4$ nanoparticles. *J. Supercrit Fluids* **2007**, *42*, 226–233. [CrossRef]

98. Goh, S.C.; Chia, C.H.; Zakaria, S.; Yusoff, M.; Haw, C.Y.; Ahmadi, S.; Huang, N.M.; Lim, H.N. Hydrothermal preparation of high saturation magnetization and coercivity cobalt ferrite nanocrystals without subsequent calcination. *Mat. Chim. Phys.* **2010**, *120*, 31–35. [CrossRef]

99. Jia, Z.; Ren., D.; Zhu, R. Synthesis, characterization and magnetic properties of $CoFe_2O_4$ nanorods. *Mater. Lett.* **2012**, *66*, 128–131. [CrossRef]

100. Fernandes de Madeiros, I.A.; Lopez-Moriyama, A.L.; Pereira de Souza, C. Effect of synthesis parameters on the size of cobalt ferrite crystallite. *Ceram. Int.* **2017**, *43*, 3962–3969. [CrossRef]

101. Phumying, S.; Labuayai, S.; Swatsitang, E.; Amornkitbamrung, V.; Maensiri, S. Nanocrystalline spinel ferrite (MFe_2O_4, M = Ni, Co, Mn, Mg, Zn) powders prepared by a simple aloe vera plant-extracted solution hydrothermal route. *Mater. Res. Bull.* **2013**, *48*, 2060–2065. [CrossRef]

102. Sarkar, K.; Mondal, R.; Dey, S.; Kumar, S. Cation vacancy and magnetic properties of $ZnFe_2O_4$ microspheres. *Phys. B Condens Matter.* **2020**, *583*, 412015. [CrossRef]

103. Vazquez-Vazquez, C.; Lopez-Quintela, M.A.; Bujan-Nunez, M.C.; Rivas, J. Finite size and surface effects on the magnetic properties of cobalt ferrite nanoparticles. *J. Nanopart Res.* **2011**, *13*, 1663–1676. [CrossRef]

104. Pilai, V.; Shah, D.O. Synthesis of high-coercivity cobalt ferrite particles using water-in-oil microemulsions. *J. Magn. Magn. Mater.* **1996**, *163*, 243–248. [CrossRef]

105. Vestal, C.R.; Zhang, Z.J. Synthesis and magnetic characterization of Mn and Co spinel ferrite-silica nanoparticles with tunable magnetic core. *Nano Lett.* **2003**, *3*, 1739–1743. [CrossRef]

106. Morrison, S.A.; Cahill, C.L.; Carpenter, E.E.; Calvin, S.; Swaminathan, R.; McHenry, M.; Harris, V.G. Magnetic and structural properties of nickel zinc ferrite nanoparticles synthesized at room temperature. *J. Appl. Phys.* **2004**, *95*, 6392–6395. [CrossRef]

107. Singh, C.; Jauhar, S.; Kumar, V.; Singh, J.; Singhal, S. Synthesis of zinc substituted cobalt ferrites via reverse micelle technique involving in situ template formation: A study on their structural, magnetic, optical and catalytic properties. *Mat. Chem. Phys.* **2015**, *156*, 188–197. [CrossRef]

108. Sharifi, I.; Shokrollahi, H.; Doroodmand, M.M.; Safi, R. Magnetic and structural studies on $CoFe_2O_4$ nanoparticles synthesized by co-precipitation, normal micelles and reverse micelles methods. *J. Magn. Magn. Mater.* **2012**, *324*, 1854–1861. [CrossRef]

109. Abbas, M.; Rao, P.B.; Nazrul Islam, M.; Woo Kim, K.W.; Naga, S.M.; Takahashi, M.; Kim, C. Size-controlled high magnetization $CoFe_2O_4$ nanospheres and nanocubes using rapid one-pot sonochemical technique. *Ceram. Int.* **2014**, *40*, 3269–3276. [CrossRef]

110. Kogias, G.; Tsakaloudi, V.; Van der Valk, P.; Zaspalis, V. Improvement of the properties of MnZn ferrite power cores through improvements on the microstructure of the compacts. *J. Magn. Magn. Mater.* **2012**, *324*, 235–241. [CrossRef]

111. Shanmugam, S.; Subramanian, B. Evolution of phase pure magnetic cobalt ferrite nanoparticles by varying the synthesis conditions of polyol method. *Mater. Sci. Eng. B.* **2020**, *252*, 114451. [CrossRef]

112. Ibrahim, A.M.; Abd El-Latif, M.M.; Mahmoud, M.M. Synthesis and characterization of nano-sized cobalt ferrite prepared via polyol method using conventional and microwave heating techniques. *J. Alloy. Comp.* **2010**, *506*, 201–204. [CrossRef]

113. Selima, S.S.; Khairy, M.; Mousa, M.A. Comparative studies on the impact of synthesis methods on structural, optical, magnetic and catalytic properties of $CuFe_2O_4$. *Ceram. Int.* **2019**, *45*, 6335–6540. [CrossRef]

114. Shakil, M.; Inayat, U.; Arshad, M.I.; Nabi, G.; Khalid, N.R.; Tariq, N.H.; Shahd, A.; Iqbale, M.Z. Influence of zinc and cadmium co-doping on optical and magnetic properties of cobalt ferrites. *Ceram. Int.* **2020**, *46*, 7767–7773. [CrossRef]

115. Margabandhu, M.; Sendhilnathan, S.; Senthikumar, S.; Gajalakshmi, D. Investigation of structural, morphological, magnetic properties and biomedical applications of Cu2+ substituted uncoated cobalt ferrite nanoparticles. *Braz. Arch. Biol. Technol.* **2016**, *59*, e16161046. [CrossRef]

116. Chen, Z.-H.; Sun, Y.-P.; Kang, Z.-T.; Chen, D. Preparation of $Zn_xCo_{1-x}Fe_2O_4$ nanoparticles by microwave-assisted ball milling. *Ceram. Int.* **2014**, *40*, 14687–14692. [CrossRef]
117. Balavijayalakshmi, J.; Suriyanarayanan, N.; Jayapraksah, R. Influence of copper on the magnetic properties of cobalt ferrite nano particles. *Mater. Lett.* **2012**, *81*, 52–54.
118. Jabbar, R.; Sabeeh, S.H.; Hameed, A.M. Structural, dielectric and magnetic properties of Mn+2 doped cobalt ferrite nanoparticles. *J. Magn. Magn. Mater.* **2020**, *494*, 165726. [CrossRef]
119. Majid, F.; Rauf, J.; Ata, S.; Bibi, I.; Malik, A.; Ibrahim, S.M.; Ali, A.; Iqbal, M. Synthesis and characterization of $NiFe_2O_4$ ferrite: Sol–gel and hydrothermal synthesis routes effect on magnetic, structural and dielectric characteristics. *Mater. Chem Phys.* **2021**, *258*, 123888. [CrossRef]
120. Bhame, S.D.; Joy, P.A. Enhanced strain sensitivity in magnetostrictive spinel ferrite $Co_{1-x}Zn_xFe_2O_4$. *J. Magn. Magn. Mater.* **2018**, *447*, 150–154. [CrossRef]
121. Ati, A.A.; Othaman, Z.; Samavati, A. Influence of cobalt on structural and magnetic properties of nickel ferrite nanoparticles. *J. Mol. Struct.* **2013**, *1052*, 177–182. [CrossRef]
122. Ortiz-Quinñonez, J.-L.; Pal, U.; Villanueva, M.S. Structural, magnetic, and catalytic evaluation of spinel Co, Ni, and Co−Ni ferrite nanoparticles fabricated by low-temperature solution combustion process. *ACS Omega* **2018**, *3*, 14986–15001. [CrossRef]
123. Nasrin, S.; Chowdhury, F.U.Z.; Hoque, S.M. Study of hydrodynamic size distribution and hyperthermia temperature of chitosan encapsulated zinc-substituted manganese nano ferrites suspension. *Phys. B. Cond. Mater.* **2019**, *561*, 54–63. [CrossRef]
124. Reddy, M.P.; Zhou, X.; Yann, A.; Du, S.; Huang, Q.; Mohamed, A.M.A. Low temperature hydrothermal synthesis, structural investigation and functional properties of $Co_xMn_{1-x}Fe_2O_4$ ($0 \leq x \leq$:1) nanoferrites. *Superlattices Microstruct.* **2015**, *81*, 233–242. [CrossRef]
125. Abdallah, H.K.I.; Moyo, T.; Msoni, J.Z. The effect of annealing temperature on the magnetic properties of $Mn_xCo_{1-x}Fe_2O_4$ ferrites nanoparticles. *J. Supercond. Nov. Magn.* **2012**, *25*, 2625–2630. [CrossRef]
126. Tomar, D.; Jeevanandam, P. Synthesis of cobalt ferrite nanoparticles with different morphologies via thermal decomposition approach and studies on their magnetic properties. *J. Alloys Comp.* **2020**, *843*, 155815. [CrossRef]
127. Rathod, S.M.; Chavan, A.R.; Jadhav, S.S.; Batoo, K.M.; Haidi, M.; Raslan, E.M. Ag+ ion substituted $CuFe_2O_4$ nanoparticles: Analysis of structural and magnetic behavior. *Chem. Phys. Lett.* **2021**, *765*, 138308. [CrossRef]
128. Manikandanath, N.T.; Rimal Isaac, R.S.; Sanjith, S.; Ramesh Kumar, P.; Anooj, E.S.; Vallinayagam, S. Superb catalytic activity of as-green synthesized copper ferrite's thermal decomposition of ammonium perchlorate. *J. Molec. Struct.* **2021**, *1234*, 130161. [CrossRef]
129. Cahyana, A.H.; Liandi, A.R.; Yulizar, Y.; Romdoni, Y.; Wendari, T.P. Green synthesis of $CuFe_2O_4$ nanoparticles mediated by Morus alba L. leaf extract: Crystal structure, grain morphology, particle size, magnetic and catalytic properties in Mannich reaction. *Ceram. Int.* **2021**. [CrossRef]
130. Sabale, S.R. Studies on catalytic activity of $MnFe_2O_4$ and $CoFe_2O_4$MNPsas mediators in hemoglobin based biosensor. *Mater. Today Proc.* **2020**, *23*, 139–146. [CrossRef]
131. Yadav, R.S.; Kuřitka, I.; Vilcakova, J.; Jamatia, T.; Machovsky, M.; Skoda, D.; Urbánek, P.; Masař, M.; Urbánek, M.; Kalina, L.; et al. Impact of sonochemical synthesis condition on the structural and physical properties of $MnFe_2O_4$ spinel ferrite nanoparticles. *Ultrason. Sonochem.* **2020**, *61*, 104839. [CrossRef]
132. Golsefidi, M.A.; Abbasi, F.; Abroudi, M.; Abroudi, M.; Abbasi, Z.; Sateei, A. Facile synthesis of MnFe2O4 nanoparticles and investigation of various reductant and capping agents on their size and morphology. *J. Mater. Sci. Mater. Electron.* **2017**, *28*, 1378–1385. [CrossRef]
133. Mozaffari, M.S.; Amighian, J.; Darsheshdar, J.A.E. Magnetic and structural studies of nickel-substituted cobalt ferrite nanoparticles, synthesized by the sol–gel method. *J. Magn. Magn. Mater.* **2018**, *350*, 19–22. [CrossRef]
134. Padmapriya, G.; Manikandan, A.; Krishnasamy, V.; Jaganathan, S.K.; Arul Antony, S. Spinel $Ni_xZn_{1-x}Fe_2O_4$ ($0.0 \leq x \leq 1.0$) nano-photocatalysts: Synthesis, characterization and photocatalytic degradation of methylene blue dye. *J. Mol. Struct.* **2016**, *1119*, 37–39. [CrossRef]
135. Karpova, T.; Vassiliev, V.; Vladimirova, E.; Osotov, V.; Ronkin, M.; Nosov, A. Synthesis of ultradisperse $NiFe_2O_4$ spinel by thermal decomposition of citrate precursors and its magnetic properties. *Ceram. Int.* **2012**, *38*, 373–379. [CrossRef]
136. Kiran Babu, L.; Rami Reddy, Y.V. A novel thermal decomposition approach for the synthesis and properties of superparamagnetic nanocrystalline $NiFe_2O_4$ and its antibacterial, electrocatalytic properties. *J. Supercond. Novel Magn.* **2020**, *33*, 1013–1021. [CrossRef]
137. Popkov, V.I.; Tolstoy, V.P.; Semenov, V.G. Synthesis of phase-pure superparamagnetic nanoparticles of $ZnFe_2O_4$ via thermal decomposition of zinc-iron layered double hydroxysulphate. *J. Alloys Comp.* **2020**, *813*, 152179. [CrossRef]
138. Sukhorukov, Y.P.; Telegin, A.V.; Bebenin, N.G.; Nosov, A.P.; Bessonov, V.D.; Buchkevich, A.A. Strain-magneto-optics of a magnetostrictive ferrimagnetic $CoFe_2O_4$. *Solid State Commun.* **2017**, *263*, 27–30. [CrossRef]
139. Salazar-Kuri, U.; Estevez, J.O.; Silva-González, N.R.; Pal, U. Large magnetostriction in chemically fabricated $CoFe_2O_4$ nanoparticles and its temperature dependence. *J. Magn. Magn. Mater.* **2018**, *460*, 141–145. [CrossRef]
140. Anantharamaiah, P.N.; Joy, P.A. Large enhancement in the magnetostriction parameters of the composite of $CoFe_2O_4$ and $CoFe_{1.9}Ga_{0.1}O_4$. *Mater. Lett.* **2019**, *236*, 303–306. [CrossRef]
141. Wang, J.; Li, J.; Li, X.; Bao, X.; Gao, X. High magnetostriction with low saturation field in highly textured $CoFe_2O_4$ by magnetic field alignment. *J. Magn. Magn. Mater.* **2018**, *462*, 53–57. [CrossRef]

142. Wang, J.; Gao, X.; Yuan, C.; Li, J.; Bao, X. Magnetostriction properties of oriented polycrystalline $CoFe_2O_4$. *J. Magn. Magn. Mater.* **2016**, *401*, 662–666. [CrossRef]

143. Li, S.; Zhang, M.; Zhan, Z.; Liu, R.; Xiong, X. Study on novel Fe-based core-shell structured soft magnetic composites with remarkable magnetic enhancement by in-situ coating nano-$ZnFe_2O_4$ layer. *J. Magn. Magn. Mater.* **2020**, *500*, 166321. [CrossRef]

144. Sinuhaji, P.; Simbolon, T.R.; Hamid, M.; Hutajulu, D.A.; Sembiring, T.; Rianna, M.; Ginting, M. Influences of Co compositions in $CoFe_2O_4$ on microstructures, thermal, and magnetic properties. *Case Stud. Therm. Eng.* **2021**, *26*, 101040. [CrossRef]

145. Revathi, J.; Abel, M.J.; Archana, V.; Sumithra, T.; Thiruneelakandan, R. Synthesis and characterization of $CoFe_2O_4$ and Ni-doped $CoFe_2O_4$ nanoparticles by chemical co-precipitation technique for photo-degradation of organic dyestuffs under direct sunlight. *Phys. B Condens Matter.* **2020**, *587*, 412136. [CrossRef]

146. Nikolić, V.N.; Vasić, M.; Milić, M.M. Observation of low- and high-temperature $CuFe_2O_4$ phase at 1100 °C: The influence of Fe^{3+} ions on $CuFe_2O_4$ structural transformation. *Ceram. Int.* **2018**, *44*, 21145–21152. [CrossRef]

147. Panda, N.; Jena, A.K.; Mohapatra, S.; Rout, S.R. Copper ferrite nanoparticle-mediated N-arylation of heterocycles: A ligand-free reaction. *Tetrahedron Lett.* **2011**, *52*, 1924–1927. [CrossRef]

148. Swapna, K.; Murthy, S.N.; Nageswar, Y.V.D. Separable and reusable copper ferrite nanoparticles for cross-coupling of aryl halides with diphenyl diselenide. *Eur. J. Org. Chem.* **2011**, 1940–1946. [CrossRef]

149. Sarode, S.A.; Bhojane, J.M.; Nagarkar, J.M. An efficient magnetic copper ferrite nanoparticle: For one pot synthesis of 2-substituted benzoxazole via redox reactions. *Tetrahedron. Lett.* **2015**, *56*, 206–210. [CrossRef]

150. Dou, R.; Cheng, H.; Ma, J.; Komarneni, S. Manganese doped magnetic cobalt ferrite nanoparticles for dye degradation via a novel heterogeneous chemical catalysis. *Mat. Chem. Phys.* **2020**, *240*, 122181. [CrossRef]

151. Casbeer, E.; Sharma, V.K.; Li, X.Z. Synthesis and photocatalytic activity of ferrites under visible light: A review. *Sep. Purif. Technol.* **2012**, *87*, 1–14. [CrossRef]

152. Mahmoodi, N.M. Photocatalytic ozonation of dyes using copper ferrite nanoparticle prepared by co-precipitation method. *Desalination* **2011**, *279*, 332–337. [CrossRef]

153. Kim, H.S.; Kim, D.; Kwak, B.S.; Han, G.B.; Um, M.H.; Kang, M. Synthesis of magnetically separable core@shell structured $NiFe_2O_4$@TiO_2 nanomaterial and its use for photocatalytic hydrogen production by methanol/water splitting. *Chem. Eng. J.* **2014**, *243*, 272–279. [CrossRef]

154. Suppuraj, P.; Thirunarayanan, G.; Swaminathan, M.; Muthuvel, I. Facile Synthesis of Spinel Nanocrystalline $ZnFe_2O_4$: Enhanced Photocatalytic and Microbial Applications. *Mat. Sci. Appl. Chem.* **2017**, *34*, 5–11. [CrossRef]

155. Medeiros, P.N.; Gomes, Y.F.; Bomio, M.R.D.; Santos, I.M.G.; Silva, M.R.S.; Paskocimas, C.A.; Li, M.S.; Motta, F.V. Influence of variables on the synthesis of $CoFe_2O_4$ pigment by the complex polymerization method. *J. Adv. Ceram.* **2015**, *4*, 135–141. [CrossRef]

156. Cavalcante, P.M.T.; Dondi, M.; Guarini, G.; Raimondo, M.; Baldi, G. Colour performance of ceramic nano-pigments. *Dyes Pigm.* **2009**, *80*, 226–232. [CrossRef]

157. Nechvílová, K.; Kalendová, A. Properties of organic coatings containing pigments with surface modified with a layer of $ZnFe_2O_4$. *Adv. Sci. Technol. Res. J.* **2015**, *9*, 51–55. [CrossRef]

158. Jebeli Moeen, S.; Vaezi, M.R.; Yousefi, A.A. Chemical synthesis of nano-crystalline nickel-zinc ferrite as a magnetic pigment. *Prog. Color. Colorants Coat.* **2010**, *3*, 9–17.

159. Wantuch, A.; Kurgan, E.; Gas, P. *Numerical Analysis on Cathodic Protection of Underground Structures, Selected Issues of Electrical Engineering and Electronics*; IEEE Xplore Digital Library: New York, NY, USA, 2016; pp. 1–4.

160. Gharagozlou, M.; Ramezanzadeh, B.; Baradaran, Z. Synthesize and characterization of a novel anticorrosive cobalt ferrite nanoparticles dispersed in silica matrix ($CoFe_2O_4$-SiO_2) to improve the corrosion protection performance of epoxy coating. *Appl. Surface Sci.* **2016**, *377*, 86–98. [CrossRef]

161. Farahani, H.; Wagiran, R.; Hamidon, M.N. Humidity sensors principle, mechanism, and fabrication technologies: A comprehensive review. *Sensors* **2014**, *14*, 7881–7939. [CrossRef] [PubMed]

162. Arshaka, K.; Twomey, K.; Egan, D. A ceramic thick film humidity sensor based on MnZn ferrite. *Sensors* **2002**, *2*, 50–61. [CrossRef]

163. Atif, M.; Asghar, M.W.; Nadeem, M.; Khalid, W.; Ali, Z.; Badshah, S. Synthesis and investigation of structural, magnetic and dielectric properties of zinc substituted cobalt ferrites. *J. Phys. Chem. Solids* **2018**, *123*, 36–42. [CrossRef]

164. Wang, H.; Huang, J.; Wang, C.; Li, D.; Ding, L.; Han, Y. Immobilization of glucose oxidase using $CoFe_2O_4$/SiO_2 nanoparticles as carrier. *Appl. Surf. Sci.* **2011**, *257*, 5739–5745. [CrossRef]

165. Shojaei, A.F.; Tabatabaeian, K.; Shakeri, S.; Karimi, F. A novel 5-fluorouracile anticancer drug sensor based on $ZnFe_2O_4$ magnetic nanoparticles ionic liquids carbon paste electrode. *Sens. Actuators B Chem.* **2016**, *230*, 607–614. [CrossRef]

166. Kershi, R.M.; Aldirham, S.H. Transport and dielectric properties of nanocrystallite cobalt ferrites: Correlation with cations distribution and crystallite size. *Mat. Chem. Phys.* **2019**, *238*, 121902. [CrossRef]

167. Sathiya Priya, A.; Geetha, D.; Kavitha, N. Effect of Al substitution on the structural, electric and impedance behavior of cobalt ferrite. *Vacuum* **2019**, *160*, 453–460. [CrossRef]

168. Ahmad, S.I.; Rauf, A.; Mohammed, T.; Bahafi, A.; Kumard, D.R.; Sureshe, M.B. Dielectric, impedance, AC conductivity and low-temperature magnetic studies of Ce and Sm co-substituted nanocrystalline cobalt ferrite. *J. Magn. Magn. Mater.* **2019**, *492*, 165666. [CrossRef]

169. Mahalakshmi, S.; Jayasri, R.; Nithiyanatham, S.; Swetha, S.; Santhi, K. Magnetic interactions and dielectric behaviour of cobalt ferrite and barium titanate multiferroics nanocomposites. *Appl. Surface Sci.* **2019**, *494*, 51–56. [CrossRef]

170. Gopalan, E.V.; Joy, P.A.; Al-Omari, I.A.; Sakthi Kumar, D. On the structural, magnetic and electrical properties of sol-gel derived nanosized cobalt ferrite. *J. Alloys Comp.* **2009**, *485*, 711–717. [CrossRef]

171. Anupama, M.K.; Rudraswamy, B.; Dhananjaya, N. Investigation on impedance response and dielectric relaxation of Ni-Zn ferrites prepared by self-combustion technique. *J. Alloy. Comp.* **2017**, *706*, 554–561. [CrossRef]

172. Ewunkem, A.J.; Rodgers, L.; Campbell, D.; Staley, C.; Subedi, K.; Boyd, S.; Graves, J.L. Experimental evolution of magnetite nanoparticle resistance in Escherichia coli. *Nanomaterials* **2021**, *11*, 790. [CrossRef] [PubMed]

173. Ashour, A.H.; El-Batal, A.I.; Abdel Maksoud, M.I.A.; El-Sayyad, G.S.; Labib, S.; Abdeltwab, E.; El-Okr, M.M. Antimicrobial activity of metal-substituted cobalt ferrite nanoparticles synthesized by sol-gel technique. *Particuology* **2018**, *40*, 141–151. [CrossRef]

174. Kooti, M.; Gharineh, S.; Mehrkhah, M.; Shaker, A.; Motamedi, H. Preparation and antibacterial activity of $CoFe_2O_4/SiO_2/Ag$ composite impregnated with streptomycin. *Chem. Eng. J.* **2015**, *259*, 34–42. [CrossRef]

175. Ramanavičius, S.; Žalneravičius, R.; Niaura, G.; Drabavičius, A.; Jagminas, A. Shell-dependent antimicrobial efficiency of cobalt ferrite nanoparticles. *Nano Struct. Nano Objects* **2018**, *15*, 40–47. [CrossRef]

176. Ishaq, K.; Saka, A.A.; Kamardeen, A.O.; Ahmed, A.; Alhassan, M.I.H.; Abdullahi, H. Characterization and antibacterial activity of nickel ferrite doped-alumina nanoparticle. *Eng. Sci. Technol.* **2017**, *20*, 563–569. [CrossRef]

177. Gas, P.; Miaskowski, A. Specifying the ferrofluid parameters important from the viewpoint of magnetic fluid hyperthermia. In Proceedings of the 2015 Selected Problems of Electrical Engineering and Electronics (WZEE), Kielce, Poland, 17–19 September 2015; pp. 1–6.

178. Suleman, M.; Riaz, S. In silico study of hyperthermia treatment of liver cancer using core-shell $CoFe_2O_4@MnFe_2O_4$ magnetic nanoparticles. *J. Magn. Magn. Mater.* **2020**, *498*, 166143. [CrossRef]

179. Issa, B.; Obaidat, I.M.; Albiss, B.A.; Haik, Y. Nanoparticles: Surface effects and properties related to biomedicine applications. *Int. J. Mol. Sci.* **2013**, *14*, 21266–21305. [CrossRef] [PubMed]

180. Kurgan, E.; Gas, P. Magnetophoretic placement of ferromagnetic nanoparticles in RF hyperthermia. In Proceedings of the 2017 Progress in Applied Electrical Engineering (PAEE), Koscielisko, Poland, 25–30 June 2017; pp. 1–4.

181. Manohar, A.; Geleta, D.D.; Krishnamoorthi, C.; Lee, J. Synthesis, characterization and magnetic hyperthermia properties of nearly monodisperse $CoFe_2O_4$ nanoparticles. *Ceram. Int.* **2020**, *46*, 28035–28041. [CrossRef]

182. Shah, S.A.; Hashmi, M.U.; Alam, S.; Shamim, A. Magnetic and bioactivity evaluation of ferrimagnetic $ZnFe_2O_4$ containing glass ceramics for the hyperthermia treatment of cancer. *J. Magn. Magn. Mater.* **2010**, *322*, 375–381. [CrossRef]

183. Kurgan, E.; Gas, P. Methods of calculation the magnetic forces acting on particles in magnetic fluids. In Proceedings of the 2018 Progress in Applied Electrical Engineering (PAEE), Koscielisko, Poland, 18–22 June 2018; pp. 1–5.

184. Chaibakhsh, N.; Moradi-Shoeili, M. Enzyme mimetic activities of spinel substituted nanoferrites (MFe_2O_4): A review of synthesis, mechanism and potential applications. *Mat. Sci. Eng. C* **2019**, *99*, 1424–1447. [CrossRef]

185. Cruz, M.M.; Ferreira, L.P.; Ramos, J.; Mendo, S.G.; Alves, A.F.; Godinho, M.; Carvalho, M.D. Enhanced magnetic hyperthermia of $CoFe_2O_4$ and $MnFe_2O_4$ nanoparticles. *J. Alloys Comp.* **2017**, *703*, 370–380. [CrossRef]

186. Oh, Y.; Moorthy, M.S.; Manivasagan, P.; Bharathiraja, S.; Oh, J. Magnetic hyperthermia and pH-responsive effective drug delivery to the sub-cellular level of human breast cancer cells by modified $CoFe_2O_4$ nanoparticles. *Biochimie* **2017**, *133*, 7–19. [CrossRef]

187. Kerroum, A.A.; Essyed, A.; Iacovita, C.; Baaziz, W.; Ihiawakrim, D.; Mounkachi, O.; Hamedoun, M.; Benyoussef, A.; Benaissa, M.; Ersen, O. The effect of basic pH on the elaboration of $ZnFe_2O_4$ nanoparticles by co-precipitation method: Structural, magnetic and hyperthermia characterization. *J. Magn. Magn. Mater.* **2019**, *478*, 239–246. [CrossRef]

188. Egizbek, K.; Kozlovskiy, A.L.; Ludzik, K.; Zdorovets, M.V.; Korolkov, I.V.; Marciniak, B.; Jazdzewska, M.; Chudoba, D.; Nazarova, A.; Kontek, R. Stability and cytotoxicity study of $NiFe_2O_4$ nanocomposites synthesized by co-precipitation and subsequent thermal annealing. *Ceram. Int.* **2020**, *48*, 16548–16555. [CrossRef]

189. Irfan, M.; Dogan, N.; Bingolbali, A.; Aliew, F. Synthesis and characterization of $NiFe_2O_4$ magnetic nanoparticles with different coating materials for magnetic particle imaging (MPI). *J. Magn. Magn. Mater.* **2021**. [CrossRef]

190. Gorgizadeh, M.; Behzadpour, N.; Salehi, F.; Daneshvar, F.; Dehdari Vais, R.; Nazari-Vanani, R.; Azarpira, N.; Lotfi, M.; Sattarahmady, N. A $MnFe_2O_4/C$ nanocomposite as a novel theranostic agent in MRI, sonodynamic therapy and photothermal therapy of a melanoma cancer model. *J. Alloys Comp.* **2020**, *816*, 152597. [CrossRef]

nanomaterials

Article

Effects of Process Variables on Properties of CoFe$_2$O$_4$ Nanoparticles Prepared by Solvothermal Process

Hong Diu Thi Duong [1], Dung The Nguyen [2,3] and Kyo-Seon Kim [1,*]

[1] Department of Chemical Engineering, Kangwon National University, Chuncheon 200-701, Kangwon-do, Korea; duonghongdiu97chy@gmail.com

[2] Faculty of Chemistry, University of Sciences, Vietnam National University, 19 Le Thanh Tong, Hoan Kiem, Hanoi 100000, Vietnam; nguyentd@hus.edu.vn

[3] CIRI University-Industry Cooperation Laboratory, University of Ulsan, Ulsan 44776, Korea

* Correspondence: kkyoseon@kangwon.ac.kr; Tel.: +82-33-250-6334

Abstract: Controlling the morphology and magnetic properties of CoFe$_2$O$_4$ nanoparticles is crucial for the synthesis of compatible materials for different applications. CoFe$_2$O$_4$ nanoparticles were synthesized by a solvothermal method using cobalt nitrate, iron nitrate as precursors, and oleic acid as a surfactant. The formation of CoFe$_2$O$_4$ nanoparticles was systematically observed by adjusting synthesis process conditions including reaction temperature, reaction time, and oleic acid concentration. Nearly spherical, monodispersed CoFe$_2$O$_4$ nanoparticles were formed by changing the reaction time and reaction temperature. The oleic acid-coated CoFe$_2$O$_4$ nanoparticles inhibited the growth of particle size after 1 h and, therefore, the particle size of CoFe$_2$O$_4$ nanoparticles did not change significantly as the reaction time increased. Both without and with low oleic acid concentration, the large-sized cubic CoFe$_2$O$_4$ nanoparticles showing ferromagnetic behavior were synthesized, while the small-sized CoFe$_2$O$_4$ nanoparticles with superparamagnetic properties were obtained for the oleic acid concentration higher than 0.1 M. This study will become a basis for further research in the future to prepare the high-functional CoFe$_2$O$_4$ magnetic nanoparticles by a solvothermal process, which can be applied to bio-separation, biosensors, drug delivery, magnetic hyperthermia, etc.

Keywords: cobalt ferrite nanoparticles; controlled synthesis; solvothermal method; oleic acid; magnetic properties

Citation: Duong, H.D.T.; Nguyen, D.T.; Kim, K.-S. Effects of Process Variables on Properties of CoFe$_2$O$_4$ Nanoparticles Prepared by Solvothermal Process. *Nanomaterials* 2021, 11, 3056. https://doi.org/10.3390/nano11113056

Academic Editor: Thomas Dippong

Received: 19 October 2021
Accepted: 8 November 2021
Published: 13 November 2021

Publisher's Note: MDPI stays neutral with regard to jurisdictional claims in published maps and institutional affiliations.

1. Introduction

Magnetic nanoparticles are very attractive for many applications in various fields, among which the iron oxide (Fe$_3$O$_4$) nanoparticles have been widely studied in the past decade due to their outstanding ability to capture the magnetic moment signal, high biocompatibility, and high chemical stability [1,2]. However, currently, magnetic ferrite nanoparticles have other transition metal atoms such as Ni, Cu, Mg, Zn, Co, and Mn instead of some iron atoms in the ferrite crystal lattice and have gained remarkable attention in recent years because of their improved unique physicochemical properties such as a high surface area-to-volume ratio, feasibility of surface functionalization, and excellent magnetic responses with magnetic fields and field gradients that can be widely applied to bio-separation, magnetic resonance imaging, biosensors, drug delivery, and magnetic hyperthermia [3–10]. The spinel-type ferrite nanoparticles (MFe$_2$O$_4$, where M(II) is a d-block transition metal such as Zn, Co, Mn, etc.) displayed remarkably enhanced properties. For example, Jang et al. [11] varied the amounts of Zn and Mn in the ferrite nanoparticles and found that the (Zn$_x$M$_{1-x}$)Fe$_2$O$_4$ ((M = Mn^{2+}, Fe^{2+}) nanoparticles exhibited much higher magnetism than conventional Fe$_3$O$_4$ nanoparticles, which, consequently, led to 8 to 14 times greater r2 (MRI contrast effect) values for magnetic resonance imaging and 4 times greater specific loss power (SLP) values for hyperthermia cancer cell treatments than conventional

37

magnetic nanoparticles. The MFe_2O_4 nanomaterials with specific physicochemical and magnetic properties have been synthesized over the years and have been conquering new horizons in numerous research fields, including high-density magnetic storage, catalysis, and biomedical theranostics. Among the MFe_2O_4 nanomaterials, $CoFe_2O_4$ nanoparticles are of great interest for biomedical applications because of their highest saturation magnetization and the highest SLP level compared to magnetite and manganese ferrite nanoparticles [12,13].

It has been widely reported that particle size, shape, composition, and structural defects are important factors that strongly influence the magnetic behaviors and, consequently, applications of the $CoFe_2O_4$ nanoparticles [14,15]. For example, the ferromagnetic $CoFe_2O_4$ nanoparticles have the advantages for permanent magnet applications such as magnetic recording and energy storage [16], while the superparamagnetic $CoFe_2O_4$ nanoparticles have the merits for biomedical applications such as hyperthermia treatment, drug delivery, and cancer therapy [17]. The $CoFe_2O_4$ nanoparticles could be synthesized by various methods including sol-gel combustion [18], thermal decomposition [19], co-precipitation [5], microemulsion [20], and solvothermal [21] and polyol [22] approaches. It should be emphasized that different synthesis methods or different synthesis process variables might cause large differences in the resulting magnetic properties of ferrite nanomaterials. The co-precipitation method has been known as the most convenient method to synthesize a large amount of ferrite nanoparticles at either room temperature or elevated temperature, but the synthesized nanoparticles usually exhibit a low degree of crystallinity and large polydispersity [23]. The microemulsion method has been more useful to obtain a narrower size range and more uniform physical properties of ferrite nanoparticles. However, this method generally involves the complicated steps to generate a uniform and stable emulsion system for further formation of the ferrite nanoparticles. In addition, the yield of product is relatively low and, thus, this method is not a very efficient process for scale-up [24]. The solvothermal methods have become popular and widely used to synthesize ferrite nanomaterials due to their simplicity, low cost, high potential on a large-scale fabrication, and, more importantly, high uniformity in both size and shape with excellent magnetic properties of the synthesized nanoparticles [25]. Since the chemical reactions take place in a closed one-pot system at relatively high temperature and high pressure, all process variables must be well designed and set up in advance. For example, ferromagnetic $CoFe_2O_4$ spheres with porous/hollow nanostructures were successfully synthesized through solvothermal processes [26]. The formation of such porous/hollow structures during the solvothermal processes involved the burst formation of small-size ferrite nanoparticles, subsequently assembly formation of those small-size nanoparticles, and, finally, particle growth via the Ostwald ripening process [27]. By introducing a strong surfactant like oleic acid, Jovanović et al. [28] reported that oleic acid formed the covalent bidentate with metal ions on the particle surface and a complete monolayer was formed at the critical concentration, which controlled the particle nucleation, growth, and assembly and eventually resulted in the formation of specific nanoparticle products with different sizes and shapes. Munjal et al. [14] also utilized oleic acid as a surfactant to synthesize monodispersed oleic-coated $CoFe_2O_4$ nanoparticles with high uniformity of both particle size and shape. The $CoFe_2O_4$ nanoparticles exhibited superparamagnetic characteristics due to the small-size effect and, thus, would be suitable for hyperthermia treatment. Repko et al. [29] reported that the nucleation and growth of $CoFe_2O_4$ nanoparticles could be terminated by controlling the solvent of pentanol or ethanol in the precursor solution, which facilitated the formation of smaller nanoparticles with better size distribution.

It is still a great challenge to controllably synthesize the $CoFe_2O_4$ nanoparticles of desired size, shape, and properties for proposed application, because a small difference in synthesis process' conditions might eventually cause a remarkable variation of product particle morphologies and characteristics. This will require a comprehensive study about the effects of synthesis process variables on the products. In this study, we synthesized $CoFe_2O_4$ nanoparticles in a solution system containing oleic acid, water, and ethanol by a solvothermal process. The effects of major process variables such as reaction temperature, reaction time, and oleic acid concentration on the morphologies and characteristics of $CoFe_2O_4$ nanoparticles were systematically investigated. We strongly believe that this study can be considered as a valuable protocol for the synthesis of morphology-controlled $CoFe_2O_4$ nanoparticles, because the particle morphology control is strictly required to synthesize highly uniform products with desired properties for proposed applications.

2. Materials and Methods

2.1. Chemicals

Deionized (DI) water was used to prepare all aqueous solutions. All chemicals used for this study are as follows: iron (III) nitrate nonahydrate ($Fe(NO_3)_3 \cdot 9H_2O$, $\geq 99\%$) (Daejung, Gyeonggi, Korea), cobalt (II) nitrate hexahydrate ($Co(NO_3)_2 \cdot 6H_2O$, Daejung, $\geq 98\%$), sodium hydroxide (NaOH, $\geq 96\%$)(Yakuri, Kyoto, Japan), ethanol (C_2H_6O, Daejung, $\geq 99.9\%$), 1- pentanol ($C_5H_{12}O$, $\geq 99\%$)(Junsei, Tokyo, Japan), n- hexane (C_6H_{14}, Daejung, $\geq 95\%$), and oleic acid ($C_{18}H_{34}O_2$, Daejung, extra pure).

2.2. Synthesis of Cobalt Ferrite Nanoparticles

The $CoFe_2O_4$ was synthesized solvothermally by conducting the reaction of metal oleate complexes in a mixture solution containing oleic acid, water, and ethanol. Firstly, a mixture of metal (Co^{2+}, Fe^{3+})–oleate complexes was prepared priorly by reactions of iron nitrate, cobalt nitrate with sodium hydroxide, and oleic acid in ethanol. Specifically, 10 mmol NaOH was dissolved in 2 mL distilled water and 10 mL ethanol was added, followed by a drop-by-drop addition of 3.8 mL oleic acid. The solution was vigorously stirred for 15 min and then was transferred to a Teflon autoclave cell. Another solution was prepared by dissolving 2 mmol $Fe(NO_3)_3 \cdot 9H_2O$ and 1 mmol $Co(NO_3)_2 \cdot 6H_2O$ in 18 mL DI water and stirring for 15 min and then was added drop by drop into the solution in a Teflon autoclave cell, which was stirred for 2 h by a magnetic stirrer afterwards. The autoclave cell with prepared solution was placed into an oven with controlled temperatures (120 °C, 140 °C, 160 °C, 180 °C, and 200 °C) for different processing times (1 h, 2 h, 4 h, 8 h, 12 h, and 16 h). Tap water was used to quickly cool down the autoclave cell to room temperature. The sediment product containing $CoFe_2O_4$ nanoparticles was collected by a permanent magnet, washed with hexane and then by ethanol four times. Finally, the $CoFe_2O_4$ nanoparticles were dried at 60 °C for 6 h before further characterization. The role of oleic acid was systematically investigated by varying its concentration from 0 to 1.5 M while keeping the same conditions for the other experimental variables.

2.3. Characterization

The crystal structure of the $CoFe_2O_4$ nanoparticles was analyzed by powder X-ray diffraction (XRD) using a X'Pert–PRO (PANalytical, Almelo, The Netherlands) diffractometer with Cu-Kα radiation (λ = 1.5406 Å) and a scanning speed of $10°/s$ in the 2θ range of 10°–80°. The average crystallite size (dXRD) was calculated from the full width at the half maximum of (311) peak by using the Scherrer formula [25]:

$$d = \frac{K\lambda}{\beta cos\theta} \tag{1}$$

where d, K, λ, β, and θ are the average crystalline size (nm), Scherrer constant, which has a value of 0.9, wavelength (nm), full width at half maximum of diffraction peaks, and diffraction angle (Brag's angle), respectively.

The morphology of $CoFe_2O_4$ nanoparticles was examined by using a JEM-2100F (Tokyo, Japan) transmission electron microscope TEM. The $CoFe_2O_4$ nanoparticles were first dispersed in n-hexane and washed by ultrasonic treatment for 30 min. Then, 0.01 wt% $CoFe_2O_4$ nanoparticle solution dispersed in hexane was dropped on a carbon-coated copper grid and then dried for 6 h before TEM measurement. X-ray spectroscopy (EDS) was utilized to determine the elemental composition of samples by using a S-4800 Field Emission Scanning Electron Microscope FE-SEM instrument (Hitachi, Tokyo, Japan).

Fourier-transform infrared (FTIR) spectroscopy and thermogravimetric analysis (TGA) were used to characterize the absorption of oleic acid on the surface of $CoFe_2O_4$ nanoparticles quantitatively. The FTIR spectra were recorded in the wavelength range of 500–4000 cm^{-1} by using Spectrum GX (Perkin Elmer, Waltham, MA, USA) equipped with the Frontier model. The TGA analysis was carried out by using the SDT Q600 instrument (TA instruments, DE, USA) with a heating rate of 10 °C/min in nitrogen gas for the temperature range of 25 °C–500 °C. The measured weight loss by TGA analysis was used to calculate the number of adsorbed oleic acid molecules (N) per unit surface area of $CoFe_2O_4$ nanoparticles by the following equation:

$$N = \frac{2\rho dw N_A}{3M(100 - w)} \tag{2}$$

where ρ, d, w, N_A, and M are the density of $CoFe_2O_4$ nanoparticles (=5.23 g/cm^3 [30]), average diameter of $CoFe_2O_4$ nanoparticles measured from TEM analysis, weight loss obtained from TGA results, which reflects the mass of oleic acid, Avogadro constant (=6.023 × 10^{23}), and molecular weight of oleic acid (=282.47 g/mol), respectively, and $(100 - w)$ reflects the mass of pure $CoFe_2O_4$ nanoparticles.

A PPMS-14 vibrating sample magnetometer (Quantum Design, San Diego, USA) was used for magnetization measurements with an applied magnetic field in the range of −15 to 15 kOe. The cation distribution of $CoFe_2O_4$ nanoparticles was characterized by LabRam Aramis laser Raman spectroscopy (Horiba Jobin Yvon, Irvine, USA) with a laser (533 nm) excitation source over the wavenumber range of 800–100 cm^{-1}. Cobalt ferrite is a spinel oxide with the chemical composition of $(Co_\delta^{2+}Fe_{1-\delta}^{3+})_A(Co_{1-\delta}^{2+} Fe_{1+\delta}^{3+})_BO_4$, where A is the tetrahedral site, B is the octahedral site, and δ is the cation distribution factor, which presents the fraction of tetrahedral (A) site [31]. The Co content in the tetrahedral site was calculated by using the following equation [28,32]:

$$\delta_{Raman} = \frac{I_{Co}}{2(I_{Co} + RI_{Fe})} \tag{3}$$

where I_{Co} and I_{Fe} are the intensities of A_{1g} (1) (~680 cm^{-1}) and of A_{1g} (2) (~615 cm^{-1}). $A_{1g}(1)$ and A_{1g} (2) demonstrated the stretching vibrations of the Fe–O and Co–O bonds in the tetrahedral site. Jovanović et al. [28] proposed that the R value of 0.5 can be applied for the oscillator strength of the Co–O bonds to the Fe–O bonds in the tetrahedral site of $CoFe_2O_4$ nanoparticles. In this study, we used this R value for the calculation of δ_{Raman} of our materials ($CoFe_2O_4$ nanoparticles) based on these previous works. The δ is special for the degree of inversion describing the cation distribution in the ferrite spinel structure.

3. Results

3.1. Effect of Reaction Time on the Morphology of CoFe$_2$O$_4$ Nanoparticles

Figure 1 shows the XRD patterns of $CoFe_2O_4$ nanoparticles prepared at 180 °C for different reaction times in an autoclave cell with a constant oleic acid concentration of 1 M. The diffraction peaks at 2θ = 30.22°, 35.45°, 43.16°, 53.2°, 57.12°, 62.67°, and 74.3° correspond to (111), (220), (311), (400), (422), (511), (440), and (533) planes, respectively, of the cubic spinel structure of $CoFe_2O_4$ nanoparticles (JCPDS card no. 22-1086). The $CoFe_2O_4$ crystals started to form after 1 h with the most notable growth of the diffraction peaks at 2θ = 35.45° and 62.67°, corresponding to the (311) and (440) planes, respectively. As the reaction time increased, the diffraction peaks corresponding to the (220), (400), (422), (511), and (533) planes also appeared. As the reaction time increased to 8–16 h, the intensity of those peaks increased remarkably, which means that the crystallinity of $CoFe_2O_4$ nanoparticles also increased. The average crystallite sizes were 4 nm, 4.2 nm, 4.3 nm, 5 nm, 5.6 nm, 5.8 nm, and 6 nm for the reaction time of 1 h, 2 h, 4 h, 8 h, 12 h, and 16 h, respectively, using the Scherrer formula. Figure 2 illustrates the TEM images of $CoFe_2O_4$ nanoparticles prepared at 180 °C for different reaction times. It could be clearly seen that well-separated nanoparticles with average particle sizes of 4 nm, 4.5 nm, 5 nm, 5.5 nm, 6.5 nm, and 7 nm were obtained for the reaction times of 1 h, 2 h, 4 h, 8 h, 12 h, and 16 h, respectively. The fringes in the $CoFe_2O_4$ nanoparticles were confirmed by HR-TEM. Figure 2g shows the lattice fringe with the fringe distance in single nanoparticles of 0.25 nm, which corresponds to the lattice spacing of (311) planes at 0.25 nm in the cubic spinel $CoFe_2O_4$. The oleic acid coverage was formed on the $CoFe_2O_4$ nanoparticles after 1 h of reaction time from FTIR results (Figure 3), but the oleic acid layer was not covering the total surface of $CoFe_2O_4$ nanoparticles and could not completely prevent the mass transfer to the $CoFe_2O_4$ nanoparticles. Therefore, the particle size of $CoFe_2O_4$ nanoparticles increased slightly as the reaction time increased. Table 1 shows the percentage elemental composition of the sample prepared at 180 °C for 16 h. It confirmed the presence of iron, cobalt, oxygen, and no impurities.

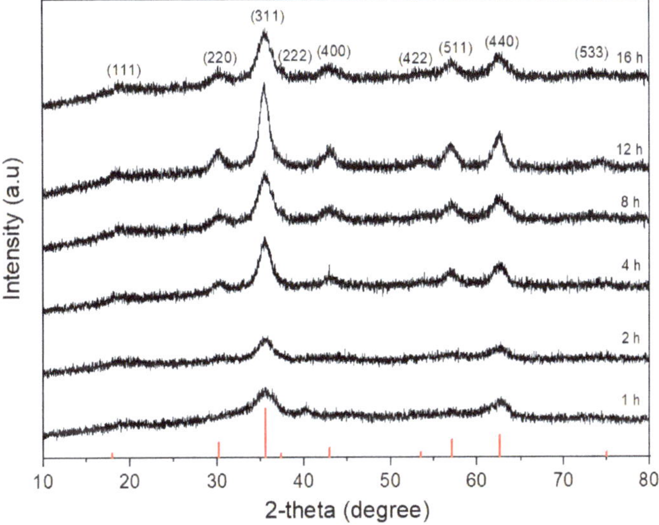

Figure 1. XRD patterns of $CoFe_2O_4$ nanoparticles synthesized at 180 °C in the presence of 1 M oleic acid for different reaction times.

FT-IR spectra were measured for the $CoFe_2O_4$ nanoparticles obtained with different reaction times, as shown in Figure 3, to investigate the effect of oleic acid coated onto the surface of nanoparticles. The strong peak at 585.2 cm^{-1} corresponded to the vibration of the Fe–O bond from the octahedral site [33,34]. This frequency band in the FT-IR spectra of all samples was associated with the characteristic peaks of the $CoFe_2O_4$ spinel structure [35]. The presence of oleic acid was confirmed by asymmetric and symmetric –CH$_2$ stretching at 2919.52 cm^{-1} and 2850 cm^{-1}, respectively. Two other bands, at 1530.4 cm^{-1} and 1406.1 cm^{-1}, could be attributed to the asymmetric and symmetric stretching vibration of the COO– group from oleic acid, respectively [36]. The peak at 3368.55 cm^{-1} was attributed to the stretching vibration of –OH, which may have been from the presence of water in the samples. The peaks at the respective wavelengths are shown in Table 2. It was observed that oleic acid was adsorbed on the surface of the $CoFe_2O_4$ nanoparticles via its carboxylate group for all the reaction times (1–16 h). Because most of the $CoFe_2O_4$ nanoparticle surfaces were covered by oleic acid from the beginning of the particle growth in the autoclave cell, the mass transfer rate of precursors from solution to nanoparticles was limited, the particle growth rate was not fast, and the $CoFe_2O_4$ nanoparticle size increased slowly with the increase of reaction time, as confirmed by the TEM analysis in Figure 2. It should be emphasized that oleic acid played an important role as a surfactant to control the particle size in the autoclave cell for different reaction times.

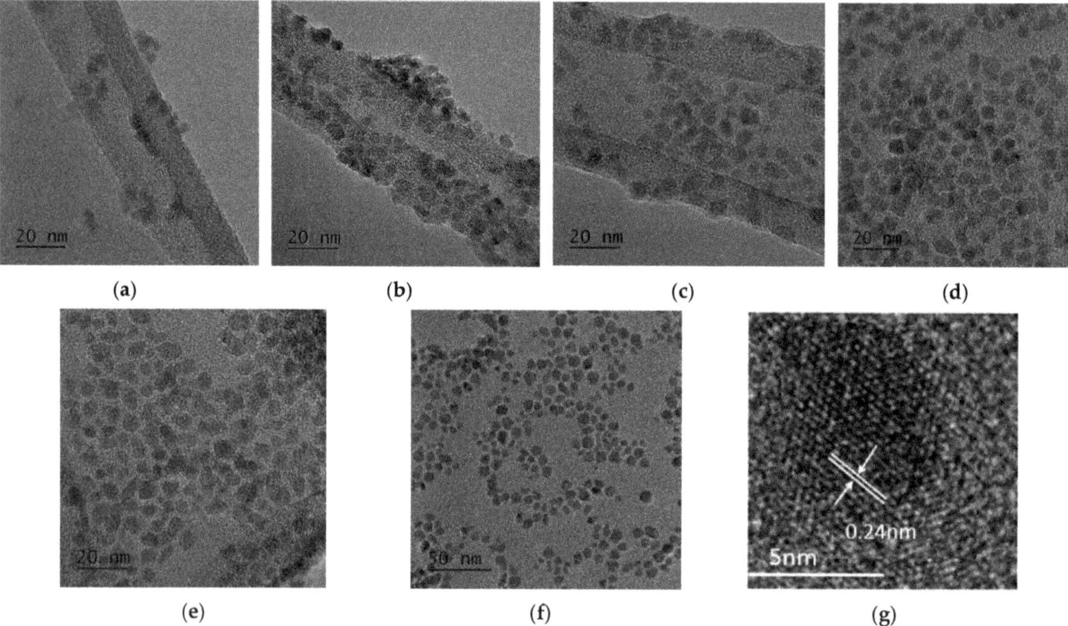

Figure 2. TEM images of CoFe$_2$O$_4$ nanoparticles synthesized at 180 °C in the presence of 1 M oleic acid for different reaction times of 1 h (**a**), 2 h (**b**), 4 h (**c**), 8 h (**d**), 12 h (**e**), and 16 h (**f**), and HR-TEM image of samples prepared for 16 h (**g**).

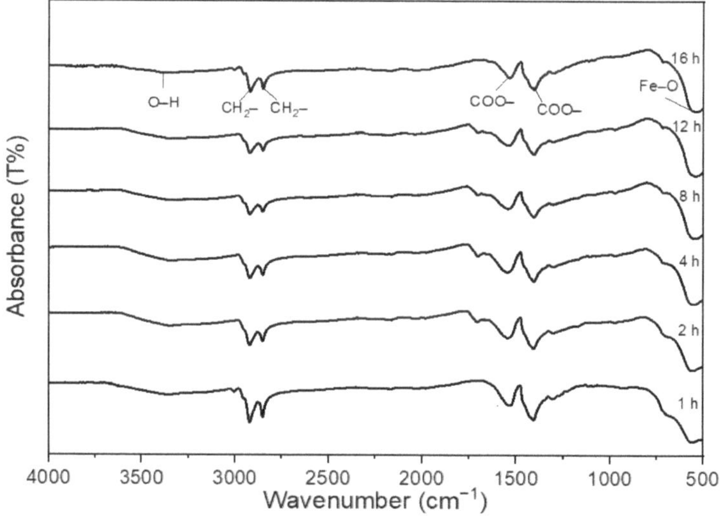

Figure 3. FT-IR spectra of CoFe$_2$O$_4$ nanoparticles synthesized at 180 °C in the presence of 1 M oleic acid for different reaction times.

Table 1. EDS for % elemental composition of $CoFe_2O_4$ nanoparticles prepared at 180 °C, 16 h.

Element	Weight (%)	Atomic (%)
O	37.93	68.46
Fe	41.3	21.36
Co	20.77	10.18
Total	100	100

Table 2. List of peaks from FT-IR results.

Peak Number	Wavelength (cm^{-1})	Functional Group
1	3368.55	O–H
2	2919.52	CH_2-
3	2850	CH_2-
4	1530.4	COO–
5	1406.1	COO–
6	585.2	Fe–O

The magnetic properties of $CoFe_2O_4$ nanoparticles prepared at 180 °C with 1 M oleic acid for different reaction times are presented in Figure 4. The M versus H dependence confirmed the superparamagnetic behavior of the prepared $CoFe_2O_4$ nanoparticles. The magnetization values (M_S) were 7.55 emu/g, 14.42 emu/g, 26.53 emu/g, 37.23 emu/g, 47.98 emu/g, and 49.35 emu/g for the reaction times of 1 h, 2 h, 4 h, 8 h, 12 h, and 16 h, respectively. The magnetization value of the $CoFe_2O_4$ nanoparticles increased quickly during the reaction time of 2–12 h, and, at 15,000 Oe of the magnetic field, increased from 7.55 emu/g to 49.35 emu/g (almost 6.5 times) with the increase of reaction time from 1 h to 16 h. This remarkable increase in the magnetic behavior of those $CoFe_2O_4$ nanoparticles was caused by the fast increases of crystallinity as well as of the particle size during the reaction time, as shown in Figures 1 and 2 [37].

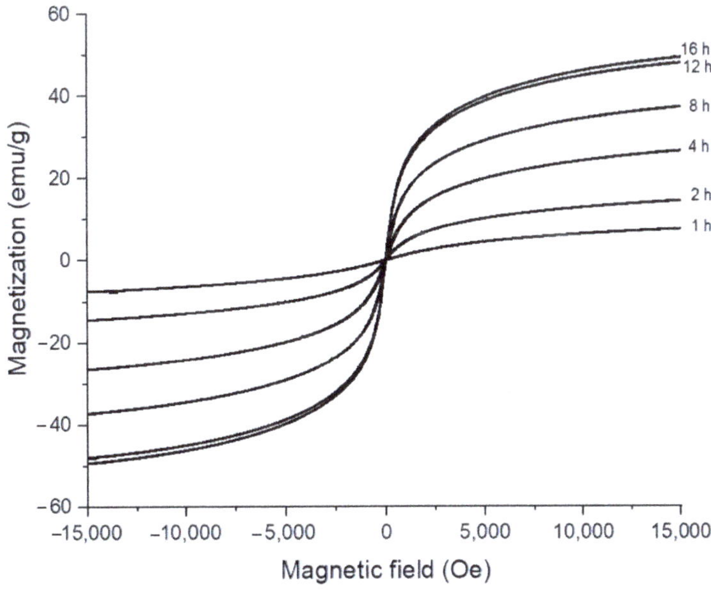

Figure 4. Room temperature M vs. H dependence of $CoFe_2O_4$ nanoparticles synthesized at 180 °C in the presence of 1 M oleic acid for different reaction times.

3.2. Effect of Reaction Temperature on the Morphology of $CoFe_2O_4$ Nanoparticles

Figure 5 shows the XRD patterns of $CoFe_2O_4$ nanoparticles prepared at different reaction temperatures with 1 M oleic acid for 16 h. The cubic spinel structure of $CoFe_2O_4$ nanoparticles was confirmed for all reaction temperatures. All the diffraction peaks became sharper as the reaction temperature increased, which shows that the crystallinity of $CoFe_2O_4$ nanoparticles increased as the reaction temperature increased. The average crystallite sizes were about 4.5 nm, 4.9 nm, 5.4 nm, 5.8 nm, 6 nm, and 9.3 nm by the Scherrer formula for the reaction temperatures of 120 °C, 140 °C, 160 °C, 180 °C, and 200 °C, respectively. The TEM images of the $CoFe_2O_4$ nanoparticles synthesized for different reaction temperatures (Figure 6) revealed that the agglomerated $CoFe_2O_4$ nanoparticles were prepared at the reaction temperature of 120 °C, while the well-separated $CoFe_2O_4$ nanoparticles were prepared at the reaction temperature of 140 °C. The average particle sizes of 5.5 nm, 6.3 nm, 7 nm, and 12 nm were obtained for the reaction temperatures of 140 °C, 160 °C, 180 °C, and 200 °C, respectively.

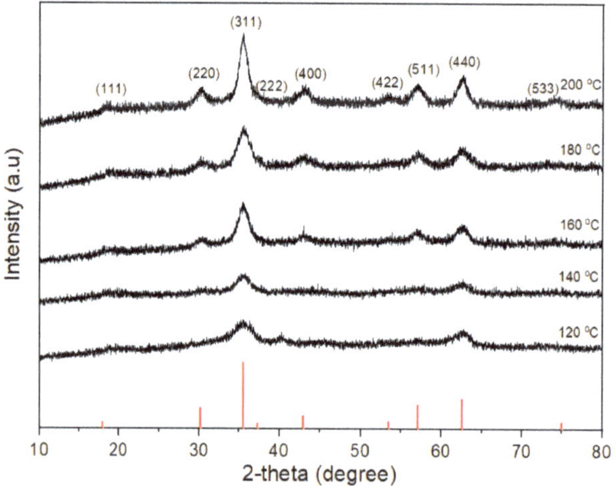

Figure 5. XRD patterns of $CoFe_2O_4$ nanoparticles synthesized at 16 h in the presence of 1 M oleic acid for different reaction temperatures.

The magnetic properties of the $CoFe_2O_4$ nanoparticles prepared for different reaction temperatures with 1 M oleic acid are presented in Figure 7. It can be observed that, with the increase of reaction temperature, the magnetization saturation of the $CoFe_2O_4$ nanoparticles increased because the crystallinity of $CoFe_2O_4$ nanoparticles also increased. The magnetization value of the $CoFe_2O_4$ nanoparticles at 15,000 Oe of magnetic field increased from 7.03 emu/g to 14.85 emu/g, 27.98 emu/g, 49.35 emu/g, and, finally, 53.3 emu/g as the reaction temperature increased from 120 °C to 140 °C, 160 °C, 180 °C, and 200 °C, respectively. The magnetization value of the $CoFe_2O_4$ nanoparticles did not increase significantly for the temperature increase from 180 °C to 200 °C because the crystallinity was already developed enough for the temperature range. The $CoFe_2O_4$ nanoparticles prepared for the reaction temperature range of 120 °C–180 °C showed superparamagnetic behavior, while the sample prepared at 200 °C showed the ferromagnetic behavior. The transition in magnetic property from superparamagnetic to ferromagnetic behavior at 200 °C is believed to be coming from the increase of particle size [38–40]. As the particle size decreased below the critical size, magnetization can randomly flip the direction under the influence of temperature, causing the residual magnetization to be 0 [41]. These results were in good agreement with previously reported values of 6–10 nm for the critical particle size of $CoFe_2O_4$ nanoparticles [5,18].

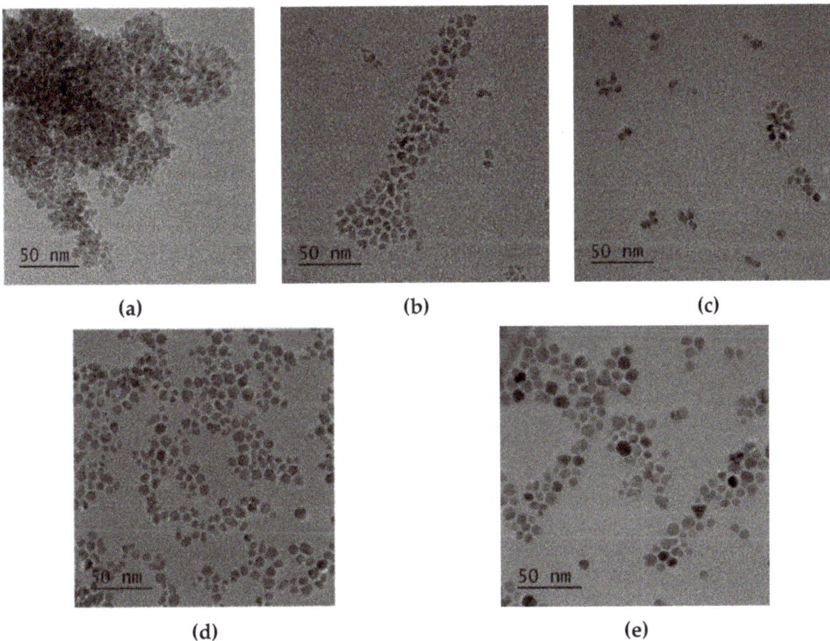

Figure 6. TEM images of cobalt ferrite nanoparticles synthesized at 16 h in the presence of 1 M oleic acid for different reaction temperatures 120 °C (**a**), 140 °C (**b**), 160 °C (**c**), 180 °C (**d**), 200 °C (**e**).

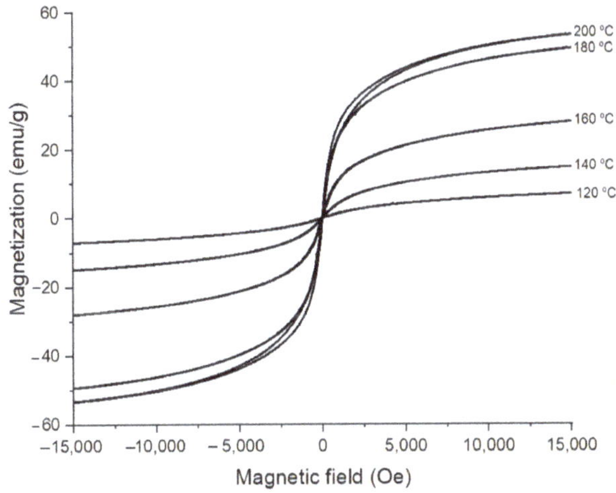

Figure 7. Room temperature M vs. H dependence of cobalt ferrite nanoparticles synthesized at 16 h in the presence of 1 M oleic acid for different reaction temperatures.

3.3. Effect of Oleic Acid Concentration on the Morphology of $CoFe_2O_4$ Nanoparticles

Figure 8 shows the TEM images of the $CoFe_2O_4$ nanoparticles prepared with different oleic acid concentrations (0 M–1.5 M) at 180 °C for 16 h. There were agglomerations between $CoFe_2O_4$ nanoparticles prepared without or with low oleic acid concentrations of 0.025 M and 0.05 M (Figure 8a–c). In these conditions, the large-sized $CoFe_2O_4$ nanoparticles of cubic shape were formed, and it was not easy to measure their independent sizes exactly because many particles were agglomerated together. For the oleic acid concentration of 0.1 M, most of the nanoparticles had the spherical shape, but a few had the large, cubic shape (Figure 8d). For the oleic acid concentration higher than 0.15 M, no cubic-shaped nanoparticle was observed. The average particle sizes of the $CoFe_2O_4$ nanoparticles prepared with oleic acid concentrations of 0.15 M, 0.5 M, 1 M, and 1.5 M were 6 nm, 6.3 nm, 7 nm, and 7.8 nm, respectively. The average particle size did not increase significantly when the oleic acid concentration was higher than 0.1 M. Figure 9 shows the XRD patterns of the $CoFe_2O_4$ nanoparticles prepared with different oleic acid concentrations (0 M–1.5 M) at 180 °C for 16 h. The cubic spinel structure of the $CoFe_2O_4$ nanoparticles was confirmed for all oleic acid concentrations (JCPDS card no. 22-1086). For the oleic acid concentrations from 0 M to 0.05 M, clear peaks including (220), (311), (400), (422), (511), (440), and (533) were presented with high intensity and the peaks of (111) and (222) were also observed with low intensity. For the oleic acid concentration of 0.1 M, all the diffraction peaks became broader and weaker with the increase of oleic acid concentration, but they became almost unchanged for the oleic acid concentration higher than 0.15 M. This can be explained by the decrease of $CoFe_2O_4$ crystallite size when the oleic acid concentration increased from 0 to 0.1 M. The average crystallite sizes with different oleic acid concentrations were found using the Scherrer formula, as shown in Figure 10. When the oleic acid concentration increased from 0.05 M to 0.1 M, the average crystallite size decreased abruptly but became almost constant for the oleic acid concentration higher than 0.1 M because the surface of the $CoFe_2O_4$ nanoparticles was almost fully covered by oleic acid and the diffusion of nanoparticle precursors from solution to $CoFe_2O_4$ nanoparticles was hindered and further growth of nanoparticles was inhibited.

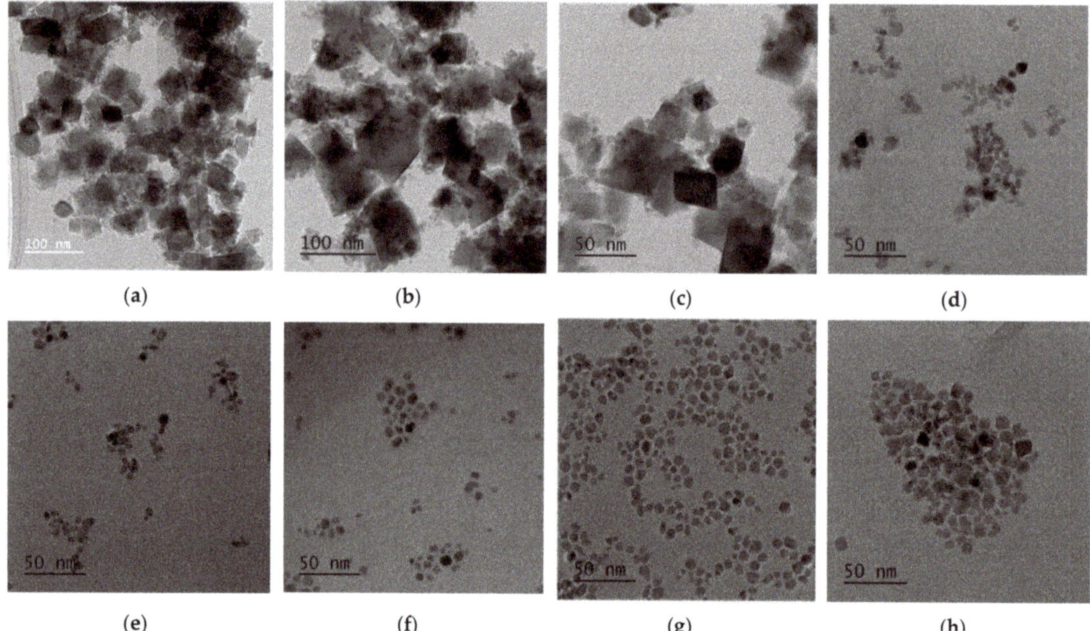

Figure 8. TEM results of cobalt ferrite nanoparticles synthesized at 180 °C for 16 h with different oleic acid concentrations of 0 M (**a**), 0.025 M (**b**), 0.05 M (**c**), 0.1 M (**d**), 0.15 M (**e**), 0.5 M (**f**), 1 M (**g**), and 1.5 M (**h**).

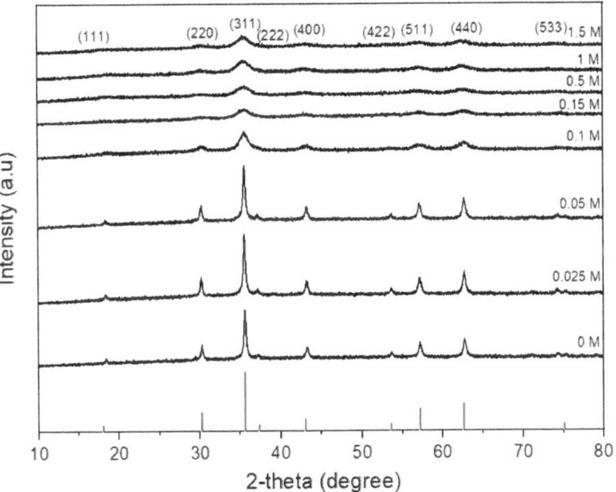

Figure 9. XRD patterns of cobalt ferrite nanoparticles synthesized at 180 °C, 16 h with different oleic acid concentrations.

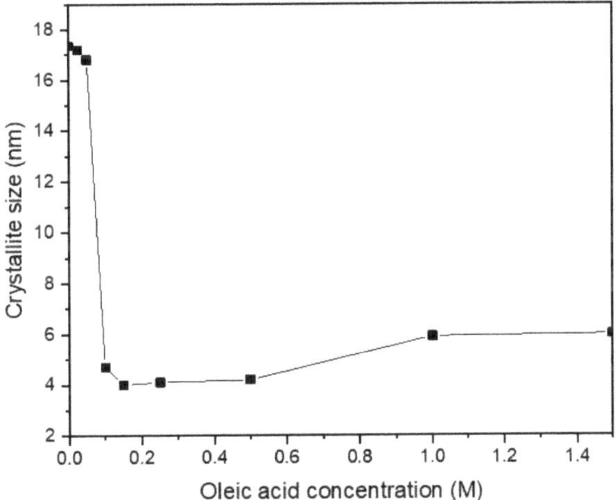

Figure 10. The average crystallite size as a function of oleic acid concentration calculated by using the Scherrer formula.

Figure 11 shows the TGA results of samples to measure the amount of oleic acid adsorbed on the surface of the $CoFe_2O_4$ nanoparticles prepared with different oleic acid concentrations. The weight loss in the temperature range of 200–450 °C can be associated with the removal of oleic acid covering the surface of the $CoFe_2O_4$ nanoparticles (the boiling point of oleic acid is 360 °C). For the oleic acid concentration above 0.1 M, the weight loss was about 21% regardless of oleic acid concentration. The numbers of oleic acid ligands per unit surface area of the $CoFe_2O_4$ nanoparticles were 2.920×10^{14}, 3.004×10^{14}, and 3.022×10^{14} with the oleic acid concentrations of 0.1 M, 0.5 M, and 1 M, respectively. These values did not change significantly for the oleic acid conditions here, which indicated that the $CoFe_2O_4$ nanoparticle surface was already fully adsorbed by the oleic acid at 0.1 M and no more adsorption was achieved for the oleic acid concentration higher than 0.1 M.

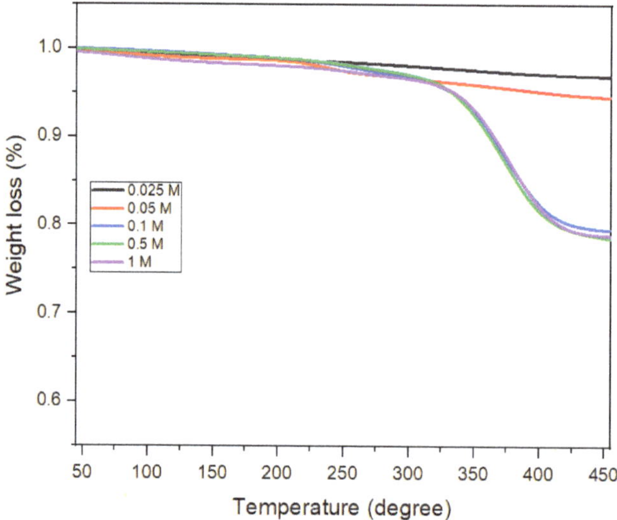

Figure 11. TGA results of cobalt ferrite nanoparticles synthesized at 180 °C, 16 h with different oleic acid concentrations.

Raman spectroscopy was used to determine the degree of cation distribution (δ) for samples prepared with different oleic acid concentrations at 180 °C, 16 h, as shown in Figure 12. The Raman spectra showed the peaks at T_{2g} (3), E_g, T_{2g} (2), A_{1g} (2), and A_{1g} (1) modes, which means that the $CoFe_2O_4$ nanoparticles of the spinel structure were synthesized. These bands, assigned as A_{1g} (1) and A_{1g} (2) modes, demonstrated the stretching vibrations of Fe–O and Co–O bonds, respectively, in the tetrahedral sites. The T_{2g} and E_g Raman modes demonstrated the vibration of the spinel structure. The Co contents in the tetrahedral site of the $CoFe_2O_4$ nanoparticles for oleic acid concentrations of 0 M, 0.05 M, 0.1 M, and 1 M were 0.303, 0.312, 0.31, and 0.32, respectively. Therefore, the oleic acid concentration had no significant influence on the cation distribution factor. The $CoFe_2O_4$ nanoparticles prepared with 1 M oleic acid concentration had the formula of $(Co_{0.32}Fe_{0.68})(Co_{0.68}Fe_{1.32})O_4$ and δ here was 0.32.

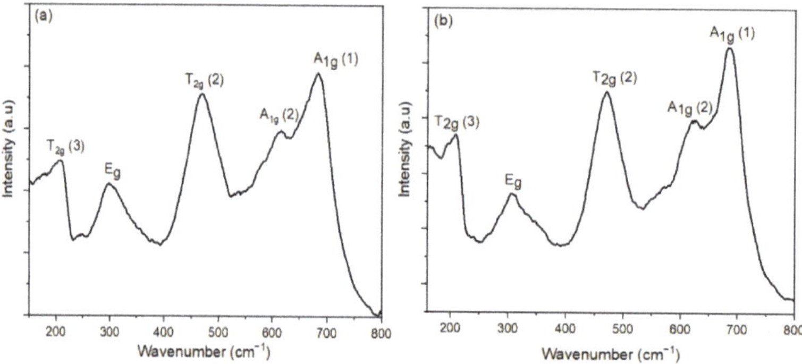

Figure 12. *Cont.*

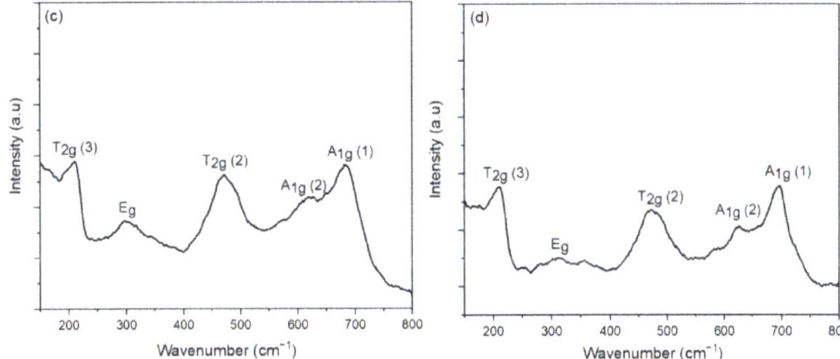

Figure 12. Raman spectra of the sample prepared with (**a**) 0 M, (**b**) 0.05 M, (**c**) 0.1 M, and (**d**) 1 M oleic acid concentration.

Magnetic properties of the $CoFe_2O_4$ nanoparticles prepared with different oleic acid concentrations were investigated by VSM measurement (Figure 13). The $CoFe_2O_4$ nanoparticles prepared with oleic acid concentrations lower than 0.05 M exhibited ferromagnetic behaviors, while the samples prepared with the oleic acid concentrations higher than 0.1 M showed the superparamagnetic behaviors. The change in magnetic behavior could be attributed to the decrease in particle size from multi-domain to single-domain structure when the oleic acid concentration reached a critical value of 0.1 M [37,42]. The formation of the single domain is detrimental to the energy, and, therefore, if thermal energy exceeds the magnetic anisotropy barrier, the residual magnetization becomes zero. The $CoFe_2O_4$ nanoparticles presented the superparamagnetic behavior with the particle size below the critical value [2]. The magnetic parameters such as saturation magnetization (M_S) and coercivity (H_C) from the hysteresis loops are listed in Table 3. The $CoFe_2O_4$ nanoparticles prepared with oleic acid showed lower magnetization saturation (M_S) than the uncoated $CoFe_2O_4$ nanoparticles prepared at the same temperature. This was due to the effect of oleic acid coating where each particle was separated from its neighbors, leading to the decrease of magnetostatic coupling between the particles [43]. The values of coercivity decreased as the oleic acid concentration increased. Coercivity depends on many factors such as surface effect, defects, strains, non-magnetic atoms, and strains in the material [44]. Thus, the decrease in coercivity was explained by interfacial defect and the decrease in agglomeration also led to the smaller coercivity [45]. The increase in the magnetic value of the samples prepared with 1 M oleic acid concentration was due to the increase in particle size.

Table 3. Saturation magnetization (M_S) and coercivity (H_C) of samples prepared with different oleic acid concentrations.

Samples	Ms (emu/g)	Hc (Oe)
0 M	54.08	4012.8
0.05 M	47.53	841.9
0.1 M	29	0
0.15 M	32	0
1 M	49	0

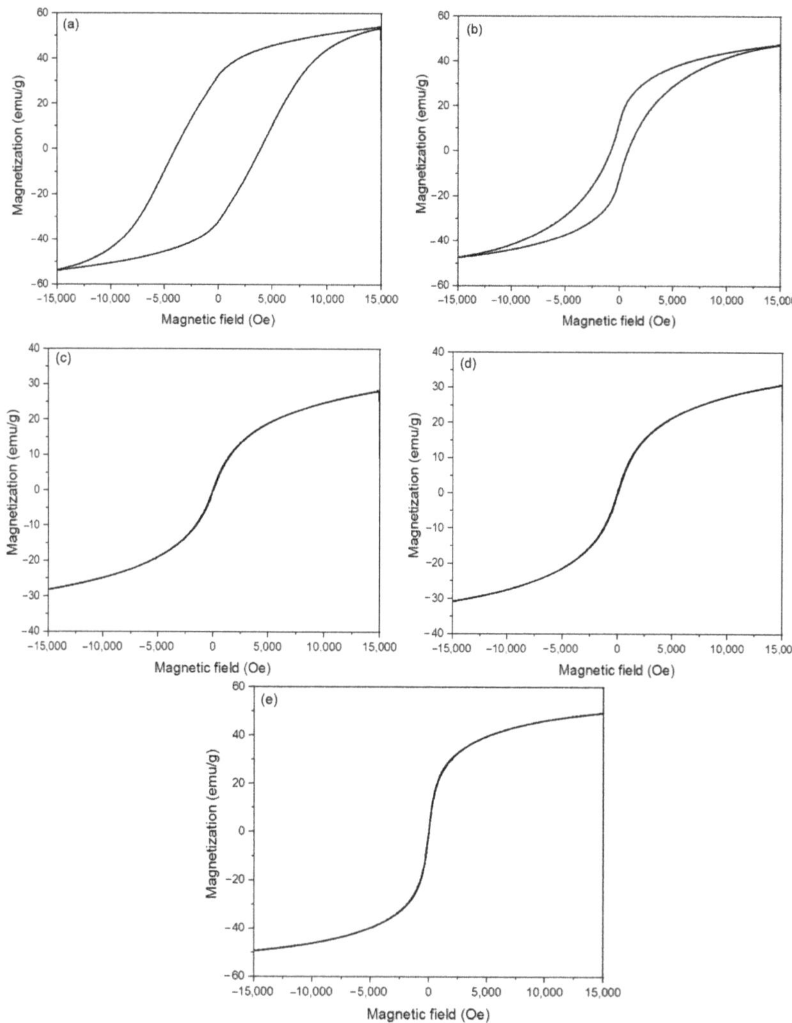

Figure 13. Room temperature M vs. H dependence of $CoFe_2O_4$ nanoparticles synthesized with different oleic acid concentrations, 0 M (**a**), 0.05 M (**b**), 0.1 M (**c**), 0.15 M (**d**), 1 M (**e**).

4. Conclusions

In this study, the $CoFe_2O_4$ magnetic nanoparticles were successfully synthesized by a solvothermal method with oleic acid as a surfactant. The effects of process variables such as reaction time, reaction temperature, and oleic acid concentration on the properties of $CoFe_2O_4$ nanoparticles were investigated. The oleic acid concentration played an important role in controlling the morphology and properties of the $CoFe_2O_4$ nanoparticles. A layer of oleic acid was coated on the surface of the $CoFe_2O_4$ nanoparticles immediately after 1 h of reaction time. This coating hindered further mass transfer of precursors from solution to nanoparticles, resulting in a negligible change in particle size with the increase of reaction time. The large-sized ferromagnetic $CoFe_2O_4$ nanoparticles with high magnetization were synthesized without or with low oleic acid concentrations. With the critical oleic acid concentration of 0.1 M, the small-sized, well-separated $CoFe_2O_4$ nanoparticles with superparamagnetic behavior were synthesized. A saturated layer of oleic acid was adsorbed on the surface of the $CoFe_2O_4$ nanoparticles when the oleic acid concentration reached a critical concentration of 0.1 M. This study will help prepare $CoFe_2O_4$ nanoparticles of high quality and also improve the performance of magnetic $CoFe_2O_4$ nanoparticles in many applications such as bio-separation, magnetic resonance imaging, biosensors, drug delivery, magnetic hyperthermia, etc.

Author Contributions: Conceptualization, H.D.T.D. and K.-S.K.; methodology, H.D.T.D.; validation, H.D.T.D.; formal analysis, H.D.T.D.; investigation, H.D.T.D., D.T.N. and K.-S.K.; resources, H.D.T.D. and K.-S.K.; data curation, H.D.T.D.; writing—original draft preparation, H.D.T.D.; writing—review and editing, D.T.N. and K.-S.K.; visualization, H.D.T.D., D.T.N. and K.-S.K.; supervision, D.T.N. and K.-S.K.; project administration, K.-S.K.; funding acquisition, K.-S.K. All authors have read and agreed to the published version of the manuscript.

Funding: This work was supported by Mid-career Researcher Program through NRF funded by the MSIP (2019R1A2C1004716).

Institutional Review Board Statement: Ethical review and approval were not applicable for this study because this research did not involve humans or animals.

Informed Consent Statement: Patient consent were not applicable for this study because this research did not involve humans.

Data Availability Statement: The data which indicated in this study are available on request from the corresponding author.

Conflicts of Interest: The authors declare no conflict of interest.

References

1. Fatima, H.; Charinpanitkul, T.; Kim, K.-S. Fundamentals to apply magnetic nanoparticles for hyperthermia therapy. *Nanomaterials* **2021**, *11*, 1203. [CrossRef]
2. Nguyen, D.T.; Kim, K.S. Controlled magnetic properties of iron oxide-based nanoparticles for smart therapy. *KONA Powder Part. J.* **2016**, *2016*, 33–47. [CrossRef]
3. Fatima, H.; Kim, K.S. Magnetic nanoparticles for bioseparation. *Korean J. Chem. Eng.* **2017**, *34*, 589–599. [CrossRef]
4. Tang, S.Q.; Moon, S.J.; Park, K.H.; Paek, S.H.; Chung, K.W.; Bae, S. Feasibility of TEOS coated $CoFe_2O_4$ nanoparticles to a GMR biosensor agent for single molecular detection. *J. Nanosci. Nanotechnol.* **2011**, *11*, 82–89. [CrossRef]
5. Gyergyek, S.; Drofenik, M.; Makovec, D. Oleic-acid-coated $CoFe_2O_4$ nanoparticles synthesized by co-precipitation and hydrothermal synthesis. *Mater. Chem. Phys.* **2012**, *133*, 515–522. [CrossRef]
6. Srinivasan, S.Y.; Paknikar, K.M.; Bodas, D.; Gajbhiye, V. Applications of cobalt ferrite nanoparticles in biomedical nanotechnology. *Nanomedicine* **2018**, *13*, 1221–1238. [CrossRef] [PubMed]
7. Gandha, K.; Elkins, K.; Poudyal, N.; Liu, J.P. Synthesis and characterization of $CoFe_2O_4$ nanoparticles with high coercivity. *J. Appl. Phys.* **2015**, *117*, 17. [CrossRef]
8. Ahmadian-Fard-Fini, S.; Ghanbari, D.; Salavati-Niasari, M. Photoluminescence carbon dot as a sensor for detecting of Pseudomonas aeruginosa bacteria: Hydrothermal synthesis of magnetic hollow $NiFe_2O_4$ -carbon dots nanocomposite material. *Compos. Part B Eng.* **2019**, *161*, 564–577. [CrossRef]
9. Joulaei, M.; Hedayati, K.; Ghanbari, D. Investigation of magnetic, mechanical and flame retardant properties of polymeric nanocomposites: Green synthesis of $MgFe_2O_4$ by lime and orange extracts. *Compos. Part B Eng.* **2019**, *176*, 107345. [CrossRef]
10. Masoumi, S.; Nabiyouni, G.; Ghanbari, D. Photo-degradation of Congored, acid brown and acid violet: Photo catalyst and magnetic investigation of $CuFe_2O_4$ –TiO_2 –Ag nanocomposites. *J. Mater. Sci. Mater. Electron.* **2016**, *27*, 11017–11033. [CrossRef]
11. Jang, J.T.; Nah, H.; Lee, J.H.; Moon, S.H.; Kim, M.G.; Cheon, J. Critical enhancements of MRI contrast and hyperthermic effects by dopant-controlled magnetic nanoparticles. *Angew. Chem. Int. Ed.* **2009**, *48*, 1234–1238. [CrossRef] [PubMed]
12. Vilar, S.Y.; Andújar, M.S.; Aguirre, C.G.; Mira, J.; Rodríguez, M.A.S.; García, S.C. A simple solvothermal synthesis of MFe_2O_4 (M=Mn, Co and Ni) nanoparticles. *J. Solid State Chem.* **2009**, *182*, 2685–2690. [CrossRef]
13. Yáñez-Vilar, S.; Sánchez-Andújar, M.; Gómez-Aguirre, C.; Mira, J.; Señarís-Rodríguez, M.A.; Castro-García, S. A simple solvothermal synthesis of MFe_2O_4 (M=Mn, Co and Ni) nanoparticles. *J. Solid State Chem.* **2009**, *182*, 2685–2690. [CrossRef]
14. Munjal, S.; Khare, N.; Sivakumar, B.; Sakthikumar, D.N. Citric acid coated $CoFe_2O_4$ nanoparticles transformed through rapid mechanochemical ligand exchange for efficient magnetic hyperthermia applications. *J. Magn. Magn. Mater.* **2019**, *477*, 388–395. [CrossRef]
15. Kazemi, M.; Ghobadi, M.; Mirzaie, A. Cobalt ferrite nanoparticles ($CoFe_2O_4$ MNPs) as catalyst and support: Magnetically recoverable nanocatalysts in organic synthesis. *Nanotechnol. Rev.* **2018**, *7*, 43–68. [CrossRef]
16. Wu, L.; Jubert, P.; Berman, D.; Imaino, W.; Nelson, A.; Zhu, H.; Zhang, S.; Sun, S. Monolayer assembly of ferrimagnetic $Co_xFe_{3-x}O_4$ nanocubes for magnetic recording. *Nano Lett.* **2014**, *14*, 3395–3399. [CrossRef] [PubMed]

17. Shen, B.; Sun, S. Chemical synthesis of magnetic nanoparticles for permanent magnet applications. *Chem.-A Eur. J.* **2020**, *26*, 6757–6766. [CrossRef] [PubMed]

18. Bortnic, R.; Goga, F.; Mesaroş, A.; Nasui, M.; Vasile, B.S.; Roxana, D.; Avram, A. Synthesis of cobalt ferrite nanoparticles via a sol-gel combustion method. *Stud. Ubb Chem.* **2016**, *4*, 213–222.

19. Tomar, D.; Jeevanandam, P. Synthesis of cobalt ferrite nanoparticles with different morphologies via thermal decomposition approach and studies on their magnetic properties. *J. Alloys Compd.* **2020**, *843*, 155815. [CrossRef]

20. Foroughi, F.; Hassanzadeh-Tabrizi, S.A.; Amighian, J. Microemulsion synthesis and magnetic properties of hydroxyapatite-encapsulated nano $CoFe_2O_4$. *J. Magn. Magn. Mater.* **2015**, *382*, 182–187. [CrossRef]

21. Kalam, A.; Al-Sehemi, A.G.; Assiri, M.; Du, G.; Ahmad, T.; Ahmada, I.; Pannipara, M. Modified solvothermal synthesis of cobalt ferrite $(CoFe_2O_4)$ magnetic nanoparticles photocatalysts for degradation of methylene blue with H_2O_2 visible light. *Results Phys.* **2018**, *8*, 1046–1053. [CrossRef]

22. Shanmugam, S.; Subramanian, B. Evolution of phase pure magnetic cobalt ferrite nanoparticles by varying the synthesis conditions of polyol method. *Mater. Sci. Eng. B Solid-State Mater. Adv. Technol.* **2020**, *252*, 114451. [CrossRef]

23. Dippong, T.; Levei, E.A.; Cadar, O. Recent advances in synthesis and applications of MFe_2O_4 (M = Co, Cu, Mn, Ni, Zn) nanoparticles. *J. Mater.* **2021**, *4*, 1560. [CrossRef] [PubMed]

24. Wu, W.; He, Q.; Jiang, C. Magnetic iron oxide nanoparticles: Synthesis and surface functionalization strategies. *Nanoscale Res. Lett.* **2008**, *3*, 397–415. [CrossRef] [PubMed]

25. Allaedini, G.; Tasirin, S.M.; Aminayi, P. Magnetic properties of cobalt ferrite synthesized by hydrothermal method. *Int. Nano Lett.* **2015**, *5*, 183–186. [CrossRef]

26. Cai, B.; Zhao, M.; Ma, Y.; Ye, Z.; Huang, J. Bioinspired formation of 3D hierarchical $CoFe_2O_4$ porous microspheres for magnetic-controlled drug release. *ACS Appl. Mater. Interfaces* **2015**, *7*, 1327–1333. [CrossRef] [PubMed]

27. Zhang, H.; Zhai, C.; Wu, J.; Ma, X.; Yang, D. Cobalt ferrite nanorings: Ostwald ripening dictated synthesis and magnetic properties. *Chem. Commun.* **2008**, *43*, 5648–5650. [CrossRef]

28. Jovanović, S.; Spreitzer, M.; Tramšek, M.; Trontelj, Z.; Suvorov, D. Effect of oleic acid concentration on the physicochemical properties of cobalt ferrite nanoparticles. *J. Phys. Chem. C* **2014**, *118*, 13844–13856. [CrossRef]

29. Repko, A.; Nižňanský, D.; Poltierová-Vejpravová, J. A study of oleic acid-based hydrothermal preparation of $CoFe_2O_4$ nanoparticles. *J. Nanoparticle Res.* **2011**, *13*, 5021–5031. [CrossRef]

30. Rajput, A.B.; Hazra, S.; Ghosh, N.N. Synthesis and characterisation of pure single-phase $CoFe_2O_4$ nanopowder via a simple aqueous solution-based EDTA-precursor route. *J. Exp. Nanosci.* **2013**, *8*, 629–639. [CrossRef]

31. Chandramohan, P.; Srinivasan, M.P.; Velmurugan, S.; Narasimhan, V.S. Cation distribution and particle size effect on Raman spectrum of $CoFe_2O_4$. *J. Solid State Chem.* **2011**, *184*, 89–96. [CrossRef]

32. Nakagomi, F.; da Silva, S.W.; Garg, V.K.; Oliveira, A.C.; Morais, P.C.; Franco, A. Influence of the Mg-content on the cation distribution in cubic $Mg_xFe_{3-x}O_4$ nanoparticles. *J. Solid State Chem.* **2009**, *182*, 2423–2429. [CrossRef]

33. Zhang, F.; Wei, C.; Wu, K.; Zhou, H.; Hu, Y.; Preis, S. Mechanistic evaluation of ferrite AFe_2O_4 (A = Co, Ni, Cu, and Zn) catalytic performance in oxalic acid ozonation. *Appl. Catal. A Gen.* **2017**, *547*, 60–68. [CrossRef]

34. Dippong, T.; Cadar, O.; Levei, E.A.; Bibicu, I.; Diamandescu, L.; Leostean, C.; Lazar, M.; Borodi, G.; Tudoran, L.B. Structure and magnetic properties of $CoFe_2O_4/SiO_2$ nanocomposites obtained by sol-gel and post annealing pathways. *Ceram. Int.* **2017**, *43*, 2113–2122. [CrossRef]

35. Stoia, M.; Stefanescu, M.; Dippong, T.; Stefanescu, O.; Barvinschi, P. Low temperature synthesis of Co_2SiO_4/SiO_2 nanocomposite using a modified sol-gel method. *J. Sol-Gel Sci. Technol.* **2010**, *54*, 49–56. [CrossRef]

36. Petcharoen, K.; Sirivat, A. Synthesis and characterization of magnetite nanoparticles via the chemical co-precipitation method. *Mater. Sci. Eng. B Solid-State Mater. Adv. Technol.* **2012**, *177*, 421–427. [CrossRef]

37. Shafi, K.V.P.M.; Gedanken, A.; Prozorov, R.; Balogh, J. Sonochemical preparation and size-dependent properties of nanostructured $CoFe_2O_4$ particles. *Chem. Mater.* **1998**, *10*, 3445–3450. [CrossRef]

38. Mohallem, N.D.S.; Seara, L.M. Magnetic nanocomposite thin films of $NiFe_2O_4/SiO_2$ prepared by sol-gel process. *Appl. Surf. Sci.* **2003**, *214*, 143–150. [CrossRef]

39. Vázquez-Vázquez, C.; Lovelle, M.; Mateo, C.; López-Quintela, M.A.; Buján-Núñez, M.C.; Serantes, D.; Baldomir, D.; Rivas, J. Maqnetocaloric effect and size-dependent study of the magnetic properties of cobalt ferrite nanoparticles prepared by solvothermal synthesis. *Phys. Status Solidi Appl. Mater. Sci.* **2008**, *205*, 1358–1362. [CrossRef]

40. Desautels, R.D.; Cadogan, J.M.; van Lierop, J. Spin dynamics in $CoFe_2O_4$ nanoparticles. *J. Appl. Phys.* **2009**, *105*, 103–106. [CrossRef]

41. Yahya, M.; Hosni, F.; Hamzaoui, A.H. Synthesis and ESR study of transition from ferromagnetism to superparamagnetism in La0.8Sr0.2MnO3 nanomanganite. *J. Intechopen* **2019**, *I*, 1–12.

42. Zhang, L.; He, R.; Gu, H. Oleic acid coating on the monodisperse magnetite nanoparticles. *Appl. Surface Sci.* **2006**, *253*, 2611–2617. [CrossRef]

43. Girgis, E.; Wahsh, M.M.; Othman, A.G.; Bandhu, L.; Rao, K. Synthesis, magnetic and optical properties of core/shell $Co_{1x}Zn_xFe_2O_4/SiO_2$ nanoparticles. *Nanoscale Res. Lett.* **2011**, *324*, 2397–2403.

44. Zubair, A.; Ahmad, Z.; Mahmood, A.; Cheong, W.; Ali, I.; Khan, M.A.; Chughtai, A.H.; Ashiq, M.N. Structural, morphological and magnetic properties of Eu-doped CoFe$_2$O$_4$ nano-ferrites. *Results Phys.* **2017**, *7*, 3203–3208. [CrossRef]
45. Suharyadi, E.; Muzakki, A.; Nofrianti, A.; Istiqomah, N.I.; Kato, T.; Iwata, S. Photocatalytic activity of magnetic core-shell CoFe$_2$O$_4$@ZnO nanoparticles for purification of methylene blue. *ACS Appl. Mater. Interfaces* **2020**, *10*, 1–6. [CrossRef]

 *nanomaterials*

Article

Sol-Gel Synthesis, Structure, Morphology and Magnetic Properties of $Ni_{0.6}Mn_{0.4}Fe_2O_4$ Nanoparticles Embedded in SiO_2 Matrix

Thomas Dippong [1,*], Erika Andrea Levei [2], Iosif Grigore Deac [3], Ioan Petean [4], Gheorghe Borodi [5] and Oana Cadar [2]

[1] Faculty of Science, Technical University of Cluj-Napoca, 76 Victoriei Street, 430122 Baia Mare, Romania
[2] INCDO-INOE 2000, Research Institute for Analytical Instrumentation, 67 Donath Street, 400293 Cluj-Napoca, Romania; erika.levei@icia.ro (E.A.L.); oana.cadar@icia.ro (O.C.)
[3] Faculty of Physics, Babes-Bolyai University, 1 Kogalniceanu Street, 400084 Cluj-Napoca, Romania; iosif.deac@phys.ubbcluj.ro
[4] Faculty of Chemistry and Chemical Engineering, Babes-Bolyai University, 11 Arany Janos Street, 400028 Cluj-Napoca, Romania; ioan.petean@ubbcluj.ro
[5] National Institute for Research and Development of Isotopic and Molecular Technologies, 65-103 Donath Street, 400293 Cluj-Napoca, Romania; borodi@itim-cj.ro
* Correspondence: thomas.dippong@cunbm.utcluj.ro

Abstract: The structure, morphology and magnetic properties of $(Ni_{0.6}Mn_{0.4}Fe_2O_4)_\alpha(SiO_2)_{100-\alpha}$ ($\alpha = 0$–100%) nanocomposites (NCs) produced by sol-gel synthesis were investigated using X-ray diffraction (XRD), Fourier transform infrared spectroscopy (FT-IR), atomic force microscopy (AFM) and vibrating sample magnetometry (VSM). At low calcination temperatures (300 °C), poorly crystallized $Ni_{0.6}Mn_{0.4}Fe_2O_4$, while at high calcination temperatures, well-crystallized $Ni_{0.6}Mn_{0.4}Fe_2O_4$ was obtained along with α-Fe_2O_3, quartz, cristobalite or iron silicate secondary phase, depending on the $Ni_{0.6}Mn_{0.4}Fe_2O_4$ content in the NCs. The average crystallite size increases from 2.6 to 74.5 nm with the increase of calcination temperature and ferrite content embedded in the SiO_2 matrix. The saturation magnetization (Ms) enhances from 2.5 to 80.5 emu/g, the remanent magnetization (M_R) from 0.68 to 12.6 emu/g and the coercive field (H_C) from 126 to 260 Oe with increasing of $Ni_{0.6}Mn_{0.4}Fe_2O_4$ content in the NCs. The SiO_2 matrix has a diamagnetic behavior with a minor ferromagnetic fraction, $Ni_{0.6}Mn_{0.4}Fe_2O_4$ embedded in SiO_2 matrix displays superparamagnetic behavior, while unembedded $Ni_{0.6}Mn_{0.4}Fe_2O_4$ has a high-quality ferromagnetic behavior.

Keywords: zinc-manganese ferrite; sol-gel; nanocomposite; magnetic properties

Citation: Dippong, T.; Levei, E.A.; Deac, I.G.; Petean, I.; Borodi, G.; Cadar, O. Sol-Gel Synthesis, Structure, Morphology and Magnetic Properties of $Ni_{0.6}Mn_{0.4}Fe_2O_4$ Nanoparticles Embedded in SiO_2 Matrix. *Nanomaterials* **2021**, *11*, 3455. https://doi.org/10.3390/nano11123455

Academic Editor: Fabien Grasset

Received: 2 December 2021
Accepted: 16 December 2021
Published: 20 December 2021

Publisher's Note: MDPI stays neutral with regard to jurisdictional claims in published maps and institutional affiliations.

1. Introduction

Nanosized mixed metal oxides with high surface area and small particle size display unique properties [1]. MFe_2O_4 (M = Zn, Co, Mn, Ni, etc.) type magnetic spinel ferrites with the general formula have numerous applications due to their high reactivity, chemical stability, optical, electrical and catalytic/ photocatalytic behaviors. Additionally, this type of magnetic nanoparticle is easily separated and recycled without important loss of their chemical activity [1,2].

Nickel ferrite ($NiFe_2O_4$) has an inverse spinel structure with Ni^{2+} ions occupying octahedral (B) sites and Fe^{3+} ions occupying tetrahedral (A) as well as octahedral (B) sites. It presents high saturation magnetization (M_S), resistivity and low losses over a large frequency range, that resulted in applications in diverse fields [3,4]. Manganese ferrite ($MnFe_2O_4$) is of great interest due to its good biocompatibility, coloristic properties, tunable magnetic properties, guidability in a magnetic field and excellent chemical stability. $MnFe_2O_4$ nanoparticles are also recognized as efficient agents for magnetic hyperthermia and magnetic resonance imaging [1–5]. $MnFe_2O_4$ has a spinel crystal structure with Fe^{3+}

ions occupying the octahedral sites and Mn^{2+} ions occupying the tetrahedral sites [4]. At calcination temperatures above 900 °C, a part of the Mn^{2+} ions migrate from tetrahedral (A) to octahedral (B) sites leading to a mixed spinel structure [4,6]. Both pure and doped $MnFe_2O_4$ tend to form anti-ferromagnetic α-Fe_2O_3 phase when are thermally treated at 200 °C in open air, but at higher calcination temperatures, the anti-ferromagnetic α-Fe_2O_3 phase is no longer remarked [7].

The substitution of $NiFe_2O_4$ with magnetic divalent transition metal ions like Mn^{2+} received considerable interest due to appealing magnetic and electrical features. Mixed Ni-Mn ferrites are frequently used, as besides the good magnetic properties, they also have large resistivity, permeability and small losses in comparison with other dielectrics [3,6]. Ni-Mn ferrites show interesting magnetic properties which recommend them to be used as hard or soft magnets and for high-frequency applications. The ferrite structure and magnetic properties are sensitive to synthesis methods, additive substitutions and calcination process [8]. By adjusting the Mn to Ni ratio in the ferrite, the magnetic properties of the ferrite can be controlled [3]. By substitution of Mn^{2+} ions with Ni^{2+} ions, Ni^{2+} ions occupy octahedral (B) sites, while Mn^{2+} ions are distributed between tetrahedral (A) and octahedral (B) sites [7].

The particle size and shape have a critical role in determining the ferrite magnetic characteristics [7]. Nanosized magnetic materials have a so-called critical particle size below which the crossover from a single- to a multidomain structure is possible. In single-domain systems, below the so-named blocking temperature, the magnetic anisotropy governs the spin alignment along the magnetization axis [7]. The presence of Mn^{2+} ions in Ni ferrites changes their structural, magnetic, electrical and dielectric properties [9]. Surface spins, spin canting and reduction of particle sizes also influence the magnetic properties [8].

The wide-scale applications of nanosized ferrites boosted the development of numerous synthesis methods. Generally, the spinel ferrites are synthesized by the ceramic technique which involves high temperatures and produces particles with low specific surface area. Alternative synthesis methods are co-precipitation, sol-gel, hydrothermal, microemulsion, heterogeneous precipitation, sonochemistry, solid-state, combustion, etc. [1–4]. Generally, the chemical methods allow the production of fine-grained ferrites, but requests a long reaction time and post-synthesis thermal treatment, and produces ferrites with poor crystallinity and broad particle size distribution [1]. Recently, the development of low-cost synthesis methods that allow the production of nano-sized, single-crystalline and single-phase powders has become of great interest [4].

The sol-gel route is an easy way to produce ferrite-based NCs (nanocomposites) as it is a simple low-cost process and concedes the control of structure and properties [5]. The sol-gel method allows the production of nanosized composite materials containing highly dispersed magnetic ferrite particles [9]. For a better control of the particle size and particle agglomeration reduction, the coating of ferrite with silica (SiO_2) is often used. The SiO_2 coating also improves the magnetic properties and biocompatibility of the ferrites due to its bio-inert behavior in contact with living tissue [5]. Most of the organic surfactants reduce the biocompatibility due to their inflammatory reactions. Oppositely, the SiO_2 is bioinert and a widely accepted material by the living body, the SiO_2 coating of ferrite nanoparticles preventing the direct contact with the living tissue and diminishing the possible inflammatory risk. Moreover, the organic surfactant layer can be removed from the nanoparticles in contact with the living tissue, revealing the ferrite surface. The SiO_2 layer cannot be solved or removed by the living tissue maintaining the optimal biocompatibility of the ferrite nanoparticles [10]. Tetraethyl orthosilicate (TEOS) is a network forming agent commonly used in the sol-gel synthesis, because it forms strong networks with moderate reactivity, permits the incorporation of various organic and inorganic molecules and offers short gelation time [5,11].

This study investigates the influence of the mixed Ni-Mn ferrite embedding in various contents of amorphous SiO_2 matrix, at different calcination temperatures on the structure, morphology and magnetic properties of $(Ni_{0.6}Mn_{04}Fe_2O_4)_\alpha(SiO_2)_{100-\alpha}$ NCs using X-ray diffraction (XRD), Fourier transform infrared spectroscopy (FT-IR), atomic force microscopy (AFM) and vibrating sample magnetometry (VSM).

2. Materials and Methods

$(Ni_{0.6}Mn_{0.4}Fe_2O_4)_\alpha(SiO_2)_{100-\alpha}$ ($\alpha = 0$–100%) NCs were obtained by the sol-gel method. Nickel nitrate $(Ni(NO_3)_2 \cdot 6H_2O)$, manganese nitrate $(Mn(NO_3)_2 \cdot 3H_2O)$ and ferric nitrate $(Fe(NO_3)_3 \cdot 9H_2O)$ were dissolved in 1,4-butanediol (1,4BD) in a molar ratio of 0.6:0.4:2:8. TEOS dissolved in ethanol and acidified with nitric acid (pH = 2) was added to the nitrate-1,4BD mixture, under continuous stirring, at room temperature, using an NO_3^-:TEOS molar ratio of 0:2 ($\alpha = 0\%$), 0.5:1.5 ($\alpha = 25\%$), 1:1 ($\alpha = 50\%$), 1.5:0.5 ($\alpha = 75\%$) and 2:0 ($\alpha = 100\%$). The obtained sol was left in open air for gelation; afterwards, the solid gels were grinded and calcined in air, at 300, 700 and 1100 °C for 5 h using an LT9 muffle furnace (Nabertherm, Lilienthal, Germany).

The structure of NCs was investigated by XRD using a D8 Advance (Bruker, Karlsruhe, Germany) diffractometer, operating at 40 kV and 40 mA and employing CuKα radiation with $\lambda = 1.54060$ Å, at room temperature. The formation of the ferrite and SiO_2 matrix were investigated using a Spectrum BX II (Perkin Elmer, Waltham, MA, USA) Fourier-transform infrared spectrometer in the range of 400–4000 cm^{-1}. AFM was carried-out using a JSPM 4210 (JEOL, Tokio, Japan) scanning probe microscope using NSC15 silicon nitride cantilevers with resonant frequency of 325 kHz and force constant of 40 N/m, in tapping mode. The NCs were dispersed into ultrapure water, transferred on glass slides by vertical adsorption for 30 s and dried in air. Several areas of variable size (2.5 μm × 2.5 μm to 1 μm × 1 μm) of the dried glass slides were scanned. A cryogenic VSM magnetometer (Cryogenic Ltd., London, UK) was used for the magnetic measurements. The M_S was determined in high magnetic field up to 10 T, whereas the magnetic hysteresis loops were conducted on samples incorporated in an epoxy resin to avoid any particle movement, between −2 to 2 T, at 300 K.

3. Results and Discussion

The XRD patterns and FT-IR spectra of the $(Ni_{0.6}Mn_{0.4}Fe_2O_4)_\alpha(SiO_2)_{100-\alpha}$ ($\alpha = 0$, 25, 50, 75, 100%) NCs calcined at 300, 700 and 1100 °C are presented in Figure 1. At all calcination temperatures, in case of NCs with $\alpha = 0\%$, the formation of amorphous SiO_2 matrix is supported by the broad halo located at $2\theta = 15$–$30°$ in the XRD pattern. At 300 °C, the NC with $\alpha = 25\%$ is amorphous, the nano-crystalline state developing by increasing the value of α (Figure 1a). In case of the NCs calcined at 700 and 1100 °C (Figure 1c,e), the observed peaks indicate the presence of the cubic spinel structure of $Mn_xNi_{1-x}Fe_2O_4$. The $MnFe_2O_4$ (JCPDS card no. 74-2403) has a lattice parameter of 8.511 Å, while the $NiFe_2O_4$ (JCPDS card no. 10-0325) [12] has a lattice parameter of 8.339 Å. The $Mn_xNi_{1-x}Fe_2O_4$ is isostructural with the two structures mentioned above, Ni and Mn being in the same position with an occupancy factor of x for Mn and $1-x$ for Ni. The reflection planes (220), (311), (222), (400), (422), (511) (440) and (533) belonging to the angular positions at $2\theta = 29.99°$, $35.33°$, $36.83°$, $42.87°$, $53.11°$, $56.65°$, $62.16°$ and $73.38°$ are consistent with the spinel structure corresponding to the Fd3m space group and match with the literature data [13]. From the positions of diffraction lines for $Mn_xNi_{1-x}Fe_2O_4$ result a lattice parameter of 8.44 Å. From the lattice parameter which has a linear dependence with x, results x = 0.6, and $Ni_{0.6}Mn_{0.4}Fe_2O_4$.

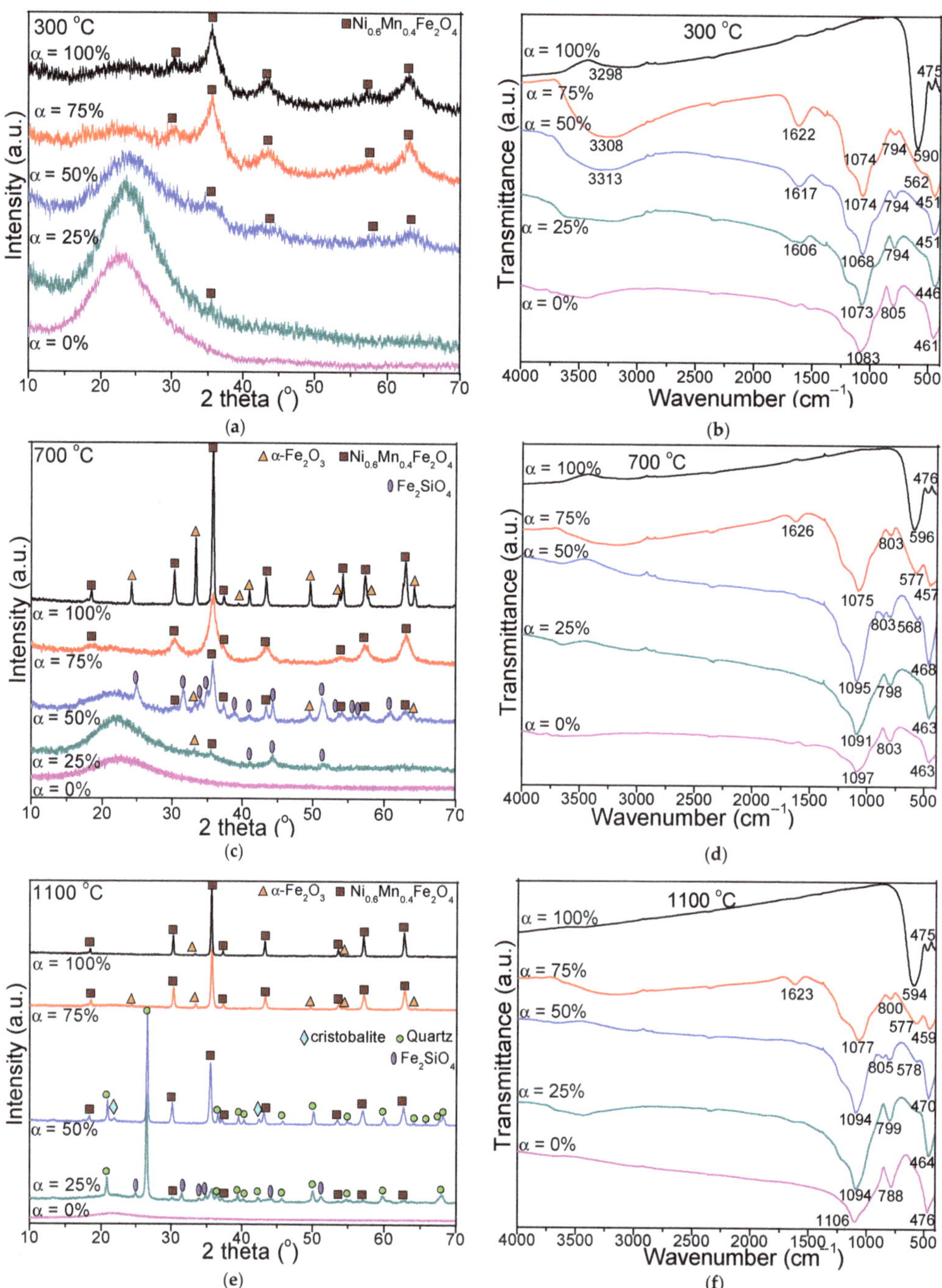

Figure 1. XRD patterns (**a,c,e**) and FT-IR spectra (**b,d,f**) of $(Ni_{0.6}Mn_{0.4}Fe_2O_4)_\alpha(SiO_2)_{100-\alpha}$ (α = 0–100%) NCs calcined at 300, 700, 1100 °C.

At 700 °C, in case of NC with $\alpha = 75\%$, the single and well-crystallized $Ni_{0.6}Mn_{04}Fe_2O_4$ is observed, while in the case of NC with $\alpha = 100\%$, the α-Fe_2O_3 (JCPDS card no. 89-0599 [12]) secondary phase is also present. The presence of α-Fe_2O_3 might be attributed to partially embedding of the ferrite in the SiO_2 matrix, due to the low content or lack of SiO_2 and the short time or calcination temperature required to produce pure crystalline $Ni_{0.6}Mn_{04}Fe_2O_4$ phase [5].

As the ferrite content decreases, in NCs with $\alpha = 25$ and 50%, besides $Ni_{0.6}Mn_{04}Fe_2O_4$, the presence of α-Fe_2O_3 and Fe_2SiO_4 (JCPDS card no. 87-0315 [12]) secondary phases is also noticed. We assume that the formation of Fe_2SiO_4 could be related to difficulty of oxygen diffusion from the pores of SiO_2 matrix and partial reduction of Fe^{3+} into Fe^{2+}, which reacts with the SiO_2 matrix and forms Fe_2SiO_4 under the reducing condition produced by the decomposition of carboxylate precursors.

At 1100 °C, in case of NCs with $\alpha = 75$–100% the well-crystallized $Ni_{0.6}Mn_{0.4}Fe_2O_4$ phase together with traces of α-Fe_2O_3 secondary phase are observed. In NCs with $\alpha = 50\%$ containing ferrite and SiO_2 matrix in 1:1 molar ratio, besides the main phase of $Ni_{0.6}Mn_{04}Fe_2O_4$, the secondary phases of crystallized SiO_2 matrix are also noticed (cristobalite, JCPDS card no. 89-8936 and quartz, JCPDS card no. 89-8936 [12]), while in NC with $\alpha = 25\%$, Fe_2SiO_4 is also obtained. Although it was reported that the thermal treatment may induce polymorphous transitions in Fe_2O_3, especially in the case of nano-sized powders or nanoparticles embedded in amorphous and porous SiO_2 matrix, in our case only α-Fe_2O_3 was observed [14]. The peaks corresponding to ferrite become more intense at 1100 °C, indicating high degree of crystallinity, crystallite size (due to the crystal coalescence process), nucleation rate and low effect of the inert surface layer [5]. Also, the highest peak shifts to higher angles with increasing $Ni_{0.6}Mn_{04}Fe_2O_4$ content embedded in the SiO_2 matrix.

Among the available methods to estimate the crystallite size, those using the diffraction profile analysis, namely Williamson-Hall and Warren-Averbach procedures, require several diffraction profiles [15,16]. Considering that in our case, especially at low calcination temperatures, we have only few diffraction peaks, we estimated the average crystallite using the Scherrer method, which requires the full the width at half maximum (FWHM) for a single diffraction line [9]. Though the X-ray profile analysis is an average method, it is still a reliable method for measuring the crystallite size, apart from transmission electron microscopy (TEM). The average crystallite size of NCs calculated using the Debye-Scherrer formula [3,17] are presented in Table 1. The low ferrite content embedded in the amorphous SiO_2 matrix retards the expansion of the crystallite size, whereas high ferrite content favors both nucleation and growth of crystallite size at the nucleation centers, leading to higher crystallite size [1]. By increasing the calcination temperature, the Ni^{2+} and Fe^{3+} ions tend to occupy specific positions in the crystal lattice of the ferrite [18,19]. The crystallites were more compact at low ferrite content embedding in SiO_2, since the smaller Ni^{2+} ion can dissolve in the spinel lattice, while high ferrite content embedding in SiO_2 matrix causes the increase of the porosity leading to higher crystallite size [18]. During the calcination process, coalescence occurs, the smaller crystallites being merged together to form the large crystallites [7].

Table 1. Structural parameters of $(N_{i0.6}Mn_{0.4}Fe_2O_4)_\alpha(SiO_2)_{100-\alpha}$ NCs calculated from AFM and XRD data.

α, %	Calcination Temperature, $^\circ$C	Roughness, nm	Average Particle Diameter, nm	Average Crystallite Size, nm	Crystallinity, %
100	300	1.0 ± 0.2	18 ± 2	4.6 ± 0.3	14 ± 1
	700	0.8 ± 0.2	52 ± 3	50 ± 3	81 ± 5
	1100	2.3 ± 0.6	75 ± 4	75 ± 5	98 ± 6
75	300	1.4 ± 0.4	20 ± 5	3.8 ± 0.3	12 ± 1
	700	1.1 ± 0.3	35 ± 3	28 ± 2	42 ± 3
	1100	2.9 ± 1.0	58 ± 5	44 ± 3	72 ± 4
50	300	1.1 ± 0.3	14 ± 1	2.6 ± 0.2	8.0 ± 0.5
	700	0.9 ± 0.2	28 ± 4	19 ± 1	25 ± 2
	1100	1.1 ± 0.4	52 ± 5	38 ± 2	66 ± 4
25	300	0.8 ± 0.2	16 ± 2	-	amorphous
	700	1.0 ± 0.3	30 ± 4	17 ± 1	21 ± 1
	1100	1.3 ± 0.3	48 ± 4	30 ± 2	56 ± 3
0	300	0.9 ± 0.2	12 ± 3	-	amorphous
	800	2.0 ± 0.8	28 ± 3	-	amorphous
	1100	2.2 ± 0.8	35 ± 4	-	amorphous

At all temperatures, the FT-IR spectra (Figure 1b,d,f) of NCs with $\alpha = 25$–100% show the absorption bands corresponding to the vibration of tetrahedral M–O (M=Ni, Mn) bonds at 568–596 cm^{-1} and of octahedral M–O (M=Fe) bonds at 446–476 cm^{-1} [1,3,4,17]. The different vibration frequencies of M–O groups are a consequence of the higher M–O bond length in octahedral (B) sites than that in tetrahedral (A) sites. The presence of these two absorption bands in FT-IR spectra confirms that the ferrites have cubic spinel structure. The intensity of the band at 568–596 cm^{-1} is larger than that of 446-476 cm^{-1}, indicating that the vibration of tetrahedral M–O is higher than of octahedral M–O groups [3]. Generally, the Ni^{2+} ions occupy the octahedral (B) sites, whereas Mn^{2+}/Fe^{3+} ions prefer both octahedral (B) and tetrahedral (A) sites [17]. The absorption bands shifting to lower values is accredited to the movement of Fe^{3+}, Mn^{2+} and Ni^{2+} ions corresponding to the O^{2-} ions in the octahedral (B) and tetrahedral (A) sites, and consequently the change of the Fe^{3+}–O^{2-} $(M^{3+}$–$O^{2-})$ and M^{2+}–O^{2-} bond length, respectively [4]. The intensity of the vibrational band at 568–596 cm^{-1} increases with the increasing calcination temperature, due to the increasing ferrite crystallinity, since the ferrites consist of crystals bonded to all adjacent neighbors through ionic, covalent or van der Waals forces [5,11,20,21].

The small shift of the vibrational band originates from the movement of ions among the tetrahedral (A) and octahedral (A) sites as a result of the increasing calcination temperature [5,11,21]. The characteristic bands of the SiO_2 matrix were detected in the FT-IR spectra of NCs with $\alpha = 0$–75%, as follows: 1068–1106 cm^{-1} with a shoulder at about 1200 cm^{-1} related to vibration of Si–O–Si chains, 788–805 cm^{-1} related to the vibrations of SiO_4 tetrahedron and 446–476 cm^{-1} related to the Si–O bond vibration and overlapping the band of Fe–O vibration [5,11]. The high intensity of these bands indicates a low polycondensation degree of the SiO_2 network [5]. The broad peaks observed at 3298-3313 cm^{-1} and at 1606–1626 cm^{-1} are ascribed to the vibrations of the –OH group and hydrogen bonds from adsorbed water molecules [1].

AFM was previously used to study the temperature effect on Ni and Mn ferrite nanoparticles transferred as thin film onto solid substrate. Ashiq et al. evidenced by AFM that Ni ferrite nanoparticles dispersion in liquid environment is proper to obtain well-structured thin films [22]. Moreover, Tong et al. reported particle diameters of 25 nm at 400 $^\circ$C; 44 nm at 500 $^\circ$C and 65 nm at 700 $^\circ$C, and surface roughness depending on the nanoparticle disposal in the topography [23].

The use of Mn ferrite nanoparticles as dispersed matter into the liquid environment as magnetic ink was also reported [24]. The printed thin film investigated with AFM revealed Mn ferrite nanoparticles of about 95 nm and the surface roughness depending on the particle diameter and on the observed agglomeration tendency [24]. The AFM topographic images are presented in Figure 2a–o. A dependence of nanoparticle diameter on the calcination temperature was observed for pure $Ni_{0.6}Mn_{04}Fe_2O_4$ (Figure 2a–c). The diameter of the round shape particles increases from about 18 nm at 300 °C to 52 nm at 700 °C, and 75 nm at 1100 °C, respectively. The crystallite size increase with the temperature increase was also observed based on the XRD data. The particle size revealed by AFM correlation with XRD crystallite size of pure Ni-Mn ferrite indicates a polycrystalline state at low temperatures (crystallite size is considerably smaller than particle size) and monocrystalline state (crystallite size is very close to the particle diameter) at 1100 °C. Establishing a certain number of crystallites per particle requires a more enhanced investigation based on scanning electron microscopy (SEM) and Brunauer–Emmett–Teller (BET) analysis [16].

XRD patterns show that the SiO_2 matrix is amorphous at all calcination temperatures. However, the particle size and shape evolution with increasing temperature may be observed using AFM. Figure 2m reveals small round shape nanoparticles and a diameter increasing with the calcination temperature, i.e., about 12 nm at 300 °C, 28 nm at 700 °C and 35 nm at 1100 °C (Figure 2n,o). Previous studies confirm the shape and sizes of the silica nanoparticles observed by AFM [25,26].

The NCs with α = 25–75% combine the morphological and structural features of both Ni-Mn ferrite and amorphous SiO_2 nanoparticles. The diameter of the round-shape nanoparticles is strongly influenced by the calcination temperature and composition (Figure 2d–l). The lowest size particles were observed at 300 °C and the bigger ones at 1100 °C (Table 1). The amorphous SiO_2 matrix increases the particle size compared to the ferrite crystallites due to the embedding effect. This effect is more visible at 300 °C than at 1100 °C. At higher calcination temperatures, the ferrite crystallite is well covered by an amorphous SiO_2 layer which forms the composite nanoparticle. The insulating behavior of the amorphous SiO_2 matrix prevents the overgrowth of magnetic domains and guarantees the nano-structural stability. A slight decrease of the nanoparticle size occurs by increasing the amorphous SiO_2 content. This decrease is most obvious at 1100 °C (Figure 2c,f,i,l), where the amorphous SiO_2 matrix inhibits the development of bigger ferrite crystallites (Table 1). A similar behavior was reported for other ferrite systems [5,11].

The powder dispersion in an aqueous environment facilitates the nanoparticle arrangement, assuring a uniform adsorption onto the solid substrate creating well-structured thin films [27], as observed in Figure 3a–o. The film roughness depends on the nanoparticle diameter and their disposal on the substrate surface. Thus, the lower roughness values are obtained at 200 °C (Figure 3a,d,g,j,m) due to the uniform adsorption of fine nanoparticles. The particle diameter increases with the calcination temperature, while the adsorbed film uniformity depends on the local heights formed by bigger nanoparticles (Figure 3c,f,i,l).

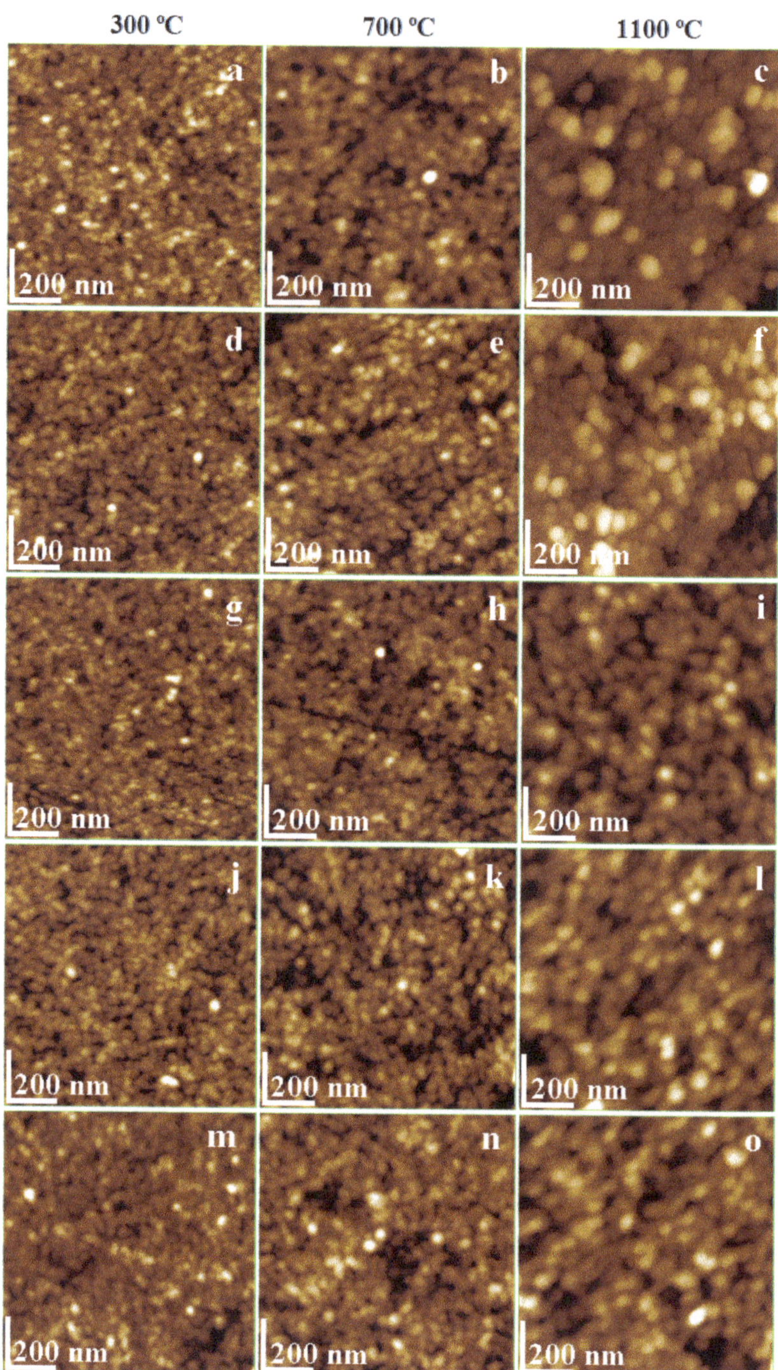

Figure 2. AFM topographic images of $(Ni_{0.6}Mn_{0.4}Fe_2O_4)_\alpha(SiO_2)_{100-\alpha}$ NCs, $\alpha 100\%$ (**a,b,c**); $\alpha = 25\%$ (**d,e,f**); $\alpha = 50\%$ (**g,h,i**); $\alpha = 75\%$ (**j,k,l**) and $\alpha = 100\%$ (**m,n,o**) calcined at 300, 700 and 1100 °C.

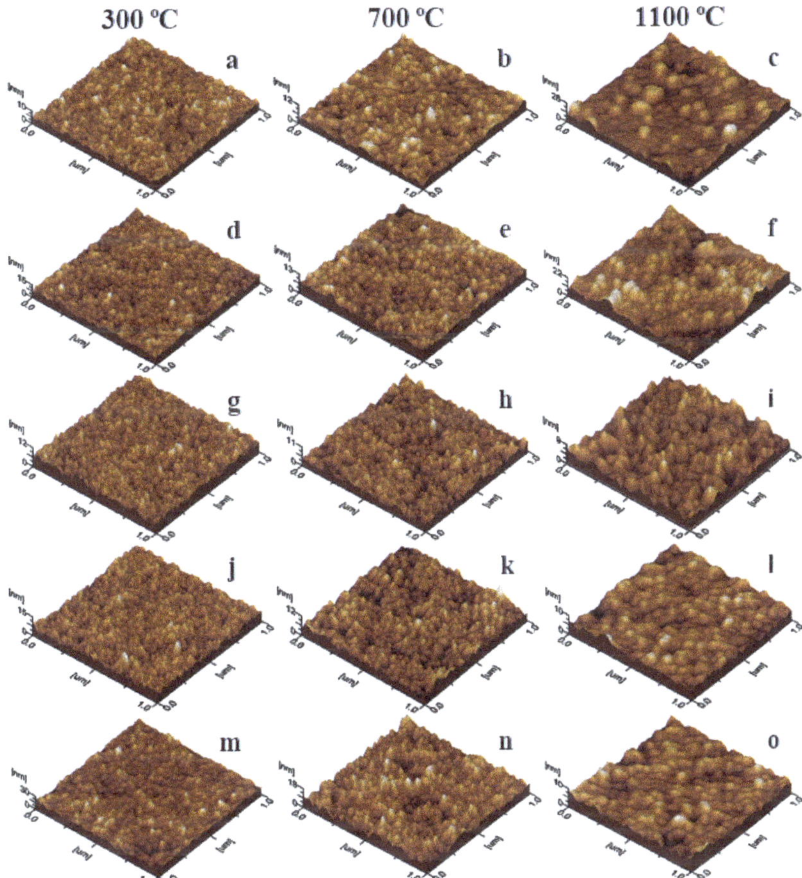

Figure 3. 3D AFM images of $(Ni_{0.6}Mn_{0.4}Fe_2O_4)_\alpha(SiO_2)_{100-\alpha}$ NCs $\alpha = 100\%$ (**a–c**); $\alpha = 25\%$ (**d–f**); $\alpha = 50\%$ (**g–i**); $\alpha = 75\%$ (**j–l**) and $\alpha = 100\%$ (**m–o**) calcined at 300, 700 and 1100 °C.

The morphological aspects of the nanoparticle thin films revealed by AFM correlated with the magnetic properties allow the design of functionalized surfaces for various applications where thermal deposition at high temperatures it is not possible, i.e., such as polymer coating.

Figures 4 and 5 display the magnetic hysteresis loops and $dM/d(\mu_0H))$ derivatives (in insets) as well as the saturation magnetization (M_S), remnant magnetization (M_R) and coercivity (H_C) values for $(Ni_{0.6}Mn_{0.4}Fe_2O_4)_\alpha(SiO_2)_{100-\alpha}$ ($\alpha = 25–100\%$) NCs calcined at 700 and 1100 °C. The hysteresis loops are very narrow, indicating that the nanoparticles have soft magnetic behavior. The derivatives of the hysteresis loops (total susceptibility) represent the local slope of M-H curves. A single sharp maximum in the $dM/d(\mu_0H)$ vs. H curves suggests the presence of a single magnetic phase.

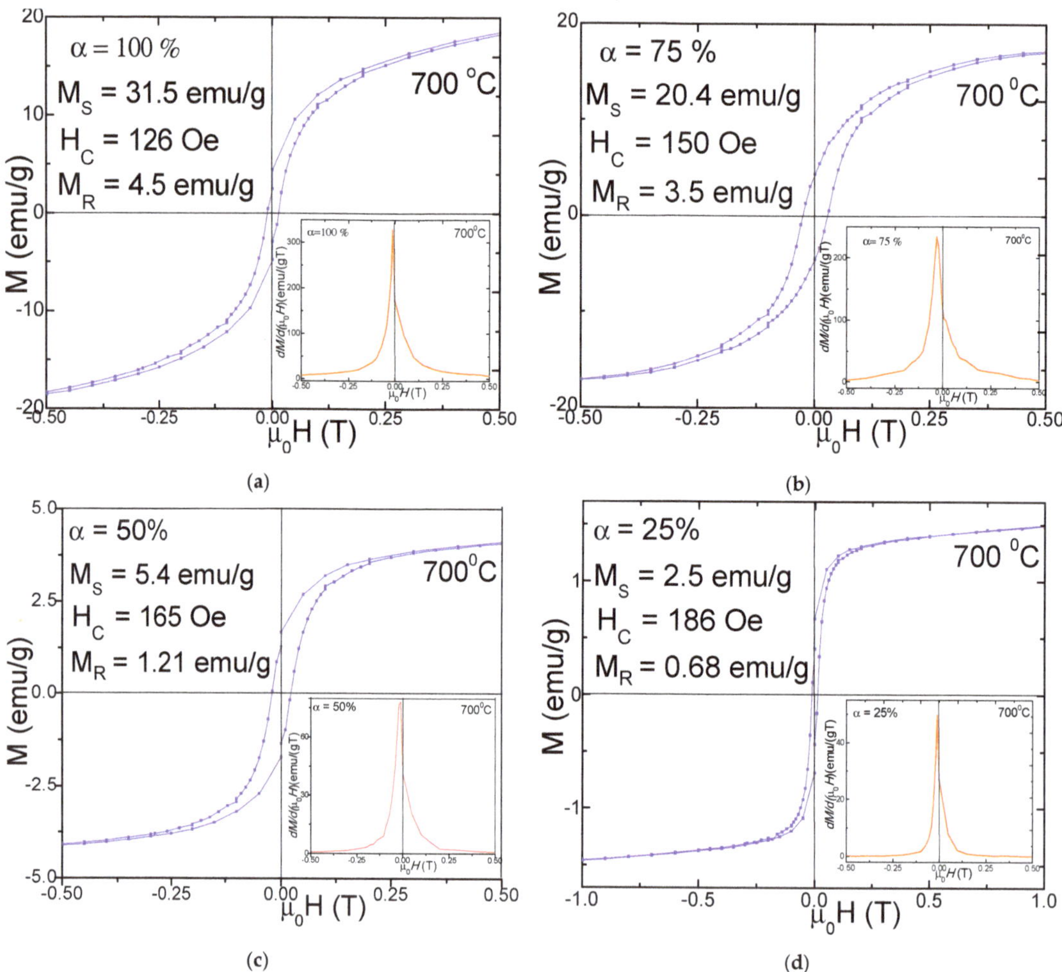

Figure 4. Magnetic hysteresis loop and magnetization derivative of $(Ni_{0.6}Mn_{0.4}Fe_2O_4)_{\alpha}(SiO_2)_{100-\alpha}$ ($\alpha = 100\%$ (**a**), 75% (**b**), 50% (**c**) and 25% (**d**)) NCs calcined at 700 °C.

The peaks' broadening indicates a larger distribution of the particle sizes. For the NCs with $\alpha = 25$–100%, $dM/d(\mu_0H)$ vs. H curves have a single and sharp peak. The morphology and the phase purity of NCs, as well as their magnetic properties, are strongly affected by the calcination temperature [3,5]. The SiO_2 matrix has diamagnetic behavior for both 700 and 1100 °C calcination temperatures. For the NCs with $\alpha = 100\%$ ($Ni_{0.6}Mn_{0.4}Fe_2O_4$), typical hysteresis loops for ferromagnetic materials were obtained, for all the calcination temperatures, due to the presence of larger size crystallites and particles as found in XRD and AFM analyses. The unembedded $Ni_{0.6}Mn_{0.4}Fe_2O_4$ ($\alpha = 100\%$) has a much higher M_S, especially when it is calcined at 1100 °C, than the ferrites embedded in the SiO_2 matrix ($\alpha = 25$–75%), with pretty narrow hysteresis loops, close to a superparamagnetic behavior.

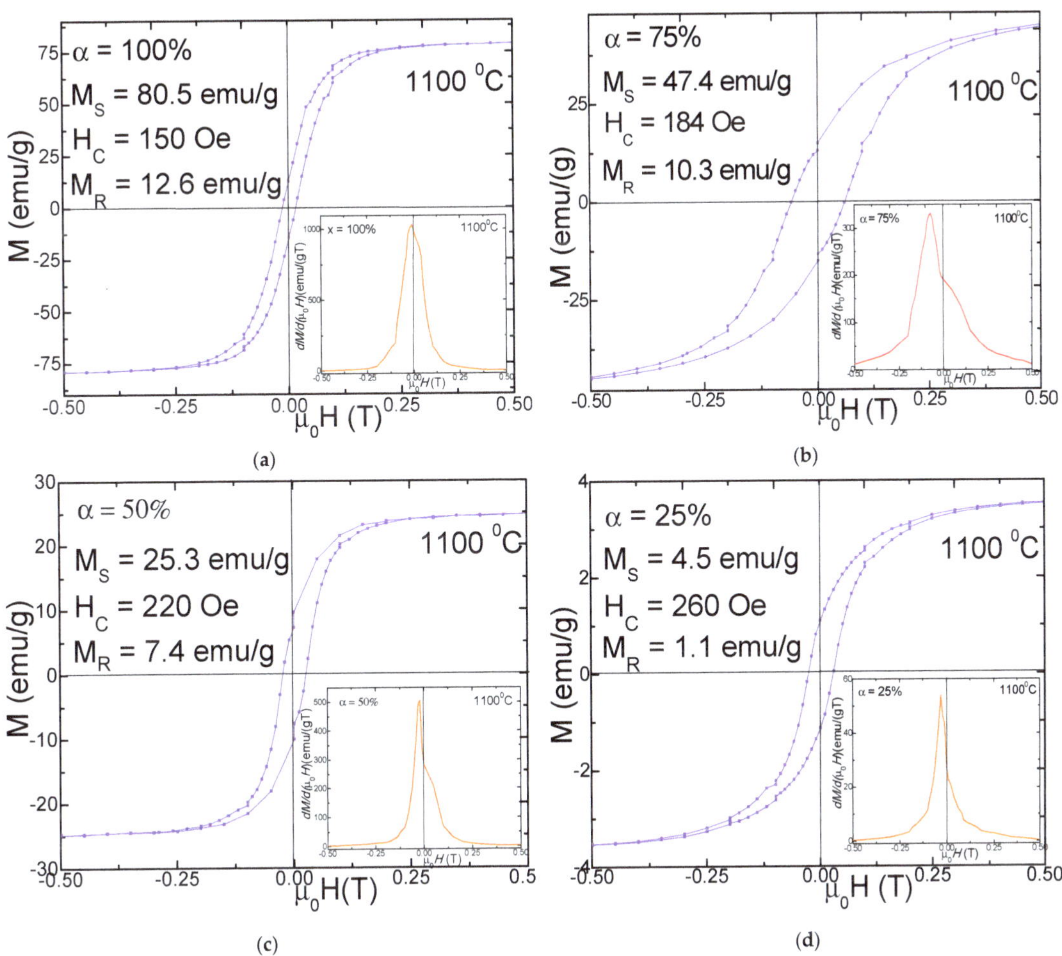

Figure 5. Magnetic hysteresis loop and magnetization derivative of $(Ni_{0.6}Mn_{0.4}Fe_2O_4)_\alpha(SiO_2)_{100-\alpha}$ ($\alpha = 100\%$ (**a**), 75% (**b**), 50% (**c**) and 25% (**d**)) NCs calcined at 1100 °C.

The superparamagnetic-like behavior of the NCs is a consequence of the low sizes of the crystallites and of their low magnetic anisotropy which allow their easily thermal activation [3,4]. The increase of the calcination temperature can lead to the improvement of M_S and M_R, as a result of a better crystallinity of the ferrite, of proper interatomic lengths changing of the atomic coordination number, etc. The M_S values of NCs with high content of $Ni_{0.6}Mn_{0.4}Fe_2O_4$ embedded in the SiO_2 matrix are larger due to larger particles sizes which show reduced spin canting and other surface effects which are usually present in small size particles. The main mechanisms of the magnetization process are the domain wall motions and the magnetic moment rotations [3]. The spin disorder on the nanoparticle surface can also strongly affect the M_S value. Moreover, the lattice defects can weaken the magnetic super-exchange interaction between the tetrahedral (A) and the octahedral (B) sites [3]. The involved magnetic Fe^{3+}, Ni^{2+} and Mn^{2+} ions have magnetic moments with the following values: 5, 2 and 5 μ_B respectively [20]. The distribution of cations between tetrahedral (A) and octahedral (B) sites of the spinel decides the magnetic moment per formula unit. The addition of Mn^{2+} ions in the Ni ferrite can induce a migration of the Fe^{3+} ions from the tetrahedral (A) to the octahedral (B) sites leading to a spin imbalance between the two sites, resulting in an increase of the magnetization at the octahedral (B) sites [20]. The surface energy of nanosized particles is large and can modify the typical cation distribution between the A and B sites [3,5]. The SiO_2 matrix can partially dilute the magnetic matrix of the cations and it can create disorder at the surface of the particles and increase the number of defects, broken bonds, canted spins, and pinning of the magnetic field lines [2,5]. The nanoparticles calcined at 700 °C have rather low values of M_S since they show lower crystallinity, large defect concentration, reduced coordination number and increased interatomic spacing [5]. The M_S values of NCs calcined at 700 °C increase with increasing $N_{i0.6}Mn_{0.4}Fe_2O_4$ content, not far from a linear dependence, from 2.5 emu/g ($\alpha = 25\%$) to 31.5 emu/g ($\alpha = 100\%$). This behavior indicates that the main contribution to magnetization is given by the ferrite content in the samples. A possible explanation of the deviation from the linear dependence can be the disorder of magnetic moments on the surface of particles, mainly for the small size particles which have a higher surface-to-volume ratio [2,5]. The increase of M_S with increasing particle sizes is typical for nano-sized ferrites [28]. Excepting the sample with $\alpha = 25\%$, there is a very good proportionality between particle and the crystallite sizes. The crystallite sizes also increase continuously with the ferrite content. This behavior suggests that the SiO_2 content has a negligible effect on the interaction between the magnetic moments of the cations from tetrahedral (A) and octahedral (B) sites, i.e., the magnetic order is not significantly changed by the SiO_2 matrix. The H_C decreases with increasing ferrite content, or with growing of the crystallite sizes as expected for multi-domain nanoparticles [28,29]. The H_C decreases from 185 Oe ($\alpha = 25\%$) to 126 Oe ($\alpha = 100\%$). The M_R decreases from 4.5 emu/g ($\alpha = 100\%$) to 0.68 emu/g ($\alpha = 25\%$) mainly due to the increasing disorder of the magnetic moments in the outer shell of the smaller sized particles [2,5]. The magnetic properties of these NCs are also affected by their bulk densities and by their grain sizes and grain size distributions. The strain released by the larger particles is higher than those of the smaller ones, resulting in lattice expansion. The pores can also contribute to the magnetic properties of the NCs, acting as pinning centers for the domain walls and for the magnetic moments of the cations [5]. The observed M_S values are in good agreement with the cation distribution theory and Neel's molecular field model [1]. The lower values of M_S for some of the NCs can be explained by the effect of the spin canting in the frame of the non-colinear Yafet-Kittel model in the presence of Jahn-Teller cations [8].

The coercive field, H_C, is given mainly by the magneto-crystalline anisotropy, but also by the exchange anisotropy due to the magnetic moment's interaction from the particle surface [15]. Generally, the M–H curves do not reach complete magnetic saturation, even in 10 T. For these cases, the M_S was estimated by using the law of approach to magnetic saturation [30,31]. The absence of complete saturation in ferromagnetic nanoparticles is generally related to the magnetic moments' disorder in the surface layers of the particles which needs a larger magnetic field for saturation, in association with the lower anisotropy of the smaller sized particles [10]. The H_C values are rather low, in the range from 126 to 260 Oe. As can be seen, the M_S increases for the NCs with lower SiO_2 matrix content. This behavior can be related to the decrease of the particle sizes with SiO_2 content increase and the associated micro-strains, and probably, the magnetic particles morphology and magnetic domain sizes [3]. The H_C decreases nearly linearly with increasing SiO_2 content due to a continuous decrease of the crystallite sizes in the single-domain range under the influence of the SiO_2 matrix [9]. The larger sized nanoparticles are composed of multi-domains, where the H_C decreases due to the formation of domain walls in the nanoparticles [7]. The measured M_S values of our previously reported $(Zn_{0.6}Mn_{0.4}Fe_2O_4)_\alpha(SiO_2)_{100-\alpha}$ ($\alpha = 100\%$) NCs [5] are similar to those belonging to the $(N_{i0.6}Mn_{0.4}Fe_2O_4)_\alpha(SiO_2)_{100-\alpha}$ ($\alpha = 100\%$) NCs. The M_S of both systems keeps the same trend, decreasing with increasing SiO_2 matrix content, which results in decreasing particle sizes. The NCs calcined at 1100 °C from the both series behave similarly showing the enhancement of the H_C with increasing SiO_2 matrix content, in spite of the much larger values of H_C for the $Zn_{0.6}Mn_{0.4}Fe_2O_4)_\alpha(SiO_2)_{100-\alpha}$ nanoparticles. These behaviors are typical for particle sizes belonging to the multi-domain range [5,28,29]. The Ni-Mn ferrites calcined at 700 °C also belong to this category, while the previous Zn-Mn ferrites calcined at 700 °C (with smaller particle sizes) behave differently, with a H_C which depreciates with decreasing SiO_2 matrix content (or with increasing particle sizes), suggesting that most of the particles have sizes belonging to single-domain range.

The obtained $(Ni_{0.6}Mn_{0.4}Fe_2O_4)_\alpha(SiO_2)_{100-\alpha}$ NCs belong to an important group of materials with potential for technical application in many biomedical and industrial fields such as drug delivery [30], hyperthermia and healthcare treatment [31,32], biocompatible magnetic fluids [33], magnetic resonance imaging contrast enhancement [34], magnetic data recording [35], microwave applications [36], supercapacitors [37] since these nanoparticles (being passivated) have low toxicity and can be operated by magnetic and electric fields [38,39].

4. Conclusions

Sol-gel route followed by calcination was used to synthesize $(Ni_{0.6}Mn_{0.4}Fe_2O_4)_\alpha(SiO_2)_{100-\alpha}$ (α = 0, 25, 50, 75, 100%) NCs. In the absence of an SiO_2 matrix (α = 100%), single-phase crystalline $Ni_{0.6}Mn_{0.4}Fe_2O_4$ was obtained at 300 °C, while at 700 and 1100 °C, ferrite is accompanied by an α-Fe_2O_3 secondary phase. By embedding high ferrite contents in the SiO_2 matrix (α = 75%), a single phase of $Ni_{0.6}Mn_{0.4}Fe_2O_4$ was obtained at 300 and 700 °C, but at 1100 °C, besides the crystalline ferrite, α-Fe_2O_3 is also present. By embedding the ferrite in equal content with the SiO_2 matrix (α = 50%), poorly crystallized single-phase $Ni_{0.6}Mn_{0.4}Fe_2O_4$ is formed at 300 °C, α-Fe_2O_3 and Fe_2SiO_4 secondary phases accompany the $Ni_{0.6}Mn_{0.4}Fe_2O_4$ at 700 °C, while at 1100 °C $Ni_{0.6}Mn_{0.4}Fe_2O_4$ is accompanied by quartz and cristobalite. The embedding of low ferrite content (α = 25%) in the SiO_2 matrix results in similar crystalline phases as in the case of NCs with α = 50% except that Fe_2SiO_4 secondary phase is also formed at 1100 °C. The increase of the calcination temperature and ferrite content embedded in the SiO_2 matrix led to an increase of the average crystallites size: 2.6–4.6 nm (300 °C), 16.5–50.1 nm (700 °C) and 30.3–74.5 nm (1100 °C). AFM investigation revealed that the average particle diameter increases with increasing calcination temperature, while the amorphous SiO_2 acts as an insulator among magnetic crystallites and prevents their overgrowth, especially at 1100 °C. The magnetic parameters enhance with increasing $Ni_{0.6}Mn_{0.4}Fe_2O_4$ content embedded in the SiO_2 matrix: M_S from 2.5 to 31.5 emu/g (700 °C) and from 4.5 to 80.5 emu/g (1100 °C), M_R from 0.68 to 4.5 emu/g (700 °C) and from 1.1 to 12.6 emu/g (1100 °C), H_C from 126 to 186 Oe (700 °C) and from 150 to 260 Oe (1100 °C). The embedding of ferrite in the SiO_2 matrix led to the particle sizes decreasing in the nano-range, but also to the alteration of the magnetic parameters. As expected, unembedded $Ni_{0.6}Mn_{0.4}Fe_2O_4$ (α = 100%) is ferromagnetic, the SiO_2 matrix (α = 0%) is diamagnetic with a small ferromagnetic fraction, while the $Ni_{0.6}Mn_{0.4}Fe_2O_4$ embedded in SiO_2 is superparamagnetic. The obtained NCs can be further developed to obtain soft and thin magnetic films on various solid substrates with tailored properties by varying the ferrite-to-matrix ratio and by a proper management of adsorption process.

Author Contributions: T.D., conceptualization, methodology, validation, writing-original draft, visualization, supervision; E.A.L., methodology, investigation (synthesis and FT-IR analysis), writing-review and editing; I.G.D., investigation (VSM analysis), writing—review and editing; I.P., investigation (AFM analysis), writing—review and editing; G.B., investigation (XRD analysis) writing-review and editing; O.C., methodology, investigation (synthesis), writing—review and editing. All authors have read and agreed to the published version of the manuscript.

Funding: This work was supported by the Romanian Ministry of Research and Innovation, CCCDI-UEFISCDI, project number PN-III-P2-2.1-PED-2019-3664 and by PNCDI III, Complex Projects of Frontier Research [PN-III-P4-ID-PCCF-2016-0112].

Institutional Review Board Statement: Not applicable.

Informed Consent Statement: Not applicable.

Data Availability Statement: Data are available from the corresponding author upon request.

Conflicts of Interest: The authors declare no conflict of interest.

References

1. Mathubala, G.; Manikandan, A.; Arul Antony, S.; Ramar, P. Photocatalytic degradation of methylene blue dye and magnetooptical studies of magnetically recyclable spinel $Ni_xMn_{1-x}Fe_2O_4$ (x = 0.0–1.0) nanoparticles. *J. Molec. Struct.* **2016**, *113*, 79–87. [CrossRef]
2. Atif, M.; Sato Turtelli, R.; Grössinger, R.; Siddique, M.; Nadeem, M. Effect of Mn substitution on the cation distribution and temperature dependence of magnetic anisotropy constant in $Co_{1-x}Mn_xFe_2O_4$ ($0.0 \leq x \leq 0.4$) ferrites. *Ceram. Int.* **2014**, *40*, 471–748. [CrossRef]
3. Suresh, J.; Trinadh, B.; Babu, B.V.; Reddy, P.V.S.S.S.N.; Mohan, B.S.; Krishna, A.R.; Samatha, K. Evaluation of micro-structural and magnetic properties of nickel nano-ferrite and Mn^{2+} substituted nickel nano-ferrite. *Phys. B Cond. Matter.* **2021**, *620*, 413264. [CrossRef]
4. Airimioaei, M.; Ciomaga, C.E.; Apostolescu, A.; Leonite, L.; Iordan, A.R.; Mitoseriu, L.; Palamaru, M.N. Synthesis and functional properties of the $Ni_{1-x}Mn_xFe_2O_4$ ferrites. *J. Alloys Comp.* **2011**, *509*, 8065–8072. [CrossRef]
5. Dippong, T.; Deac, I.G.; Cadar, O.; Levei, E.A. Effect of silica embedding on the structure, morphology and magnetic behavior of $(Zn_{0.6}Mn_{0.4}Fe_2O_4)_\delta/(SiO_2)_{(100-\delta)}$ nanoparticles. *Nanomaterials* **2021**, *11*, 2232. [CrossRef]
6. Marinca, T.F.; Chicinaș, I.; Isnard, O.; Neamțu, B.V. Nanocrystalline/nanosized manganese substituted nickel ferrites— $Ni_{1-x}Mn_xFe_2O_4$ obtained by ceramic-mechanical milling route. *Ceram. Int.* **2016**, *42*, 4754–4763. [CrossRef]
7. Maaz, K.; Duan, J.L.; Karim, S.; Chen, Y.H.; Zhai, P.F.; Xu, L.J.; Yao, H.J.; Liu, J. Fabrication and size dependent magnetic studies of $Ni_xMn_{1-x}Fe_2O_4$ (x = 0.2) cubic nanoplates. *J. Alloys Comp.* **2016**, *684*, 656–662. [CrossRef]
8. Abdallah, H.M.I.; Moyo, T. Superparamagnetic behavior of $Mn_xNi_{1-x}Fe_2O_4$ spinel nanoferrites. *J. Magn. Mater.* **2014**, *361*, 170–174. [CrossRef]
9. Shobana, M.K.; Sankar, S. Structural, thermal and magnetic properties of $Ni_{1-x}Mn_xFe_2O_4$ nanoferrites. *J. Magn. Mater.* **2009**, *321*, 2125–2128. [CrossRef]
10. Dippong, T.; Levei, E.A.; Cadar, O. Recent advances in synthesis and applications of MFe_2O_4 (M = Co, Cu, Mn, Ni, Zn) nanoparticles. *Nanomaterials* **2021**, *11*, 1560. [CrossRef]
11. Dippong, T.; Deac, I.G.; Cadar, O.; Levei, E.A.; Petean, I. Impact of Cu^{2+} substitution by Co^{2+} on the structural and magnetic properties of $CuFe_2O_4$ synthesized by sol-gel route. *Mater. Caract.* **2020**, *163*, 110248. [CrossRef]
12. Swarthmore, P. *Powder Diffraction File, Joint Committee on Powder Diffraction Standards*; International Center for Diffraction Data: Swarthmore, PA, USA, 1999.
13. Jesudoss, S.K.; Judith Vijaya, J.; John Kennedy, L.; Iyyappa Rajana, P.; Al-Lohedan, A.H.; Jothi Ramalingam, R.; Kaviyarasu, K.; Bououdina, M. Studies on the efficient dual performance of $Mn_{1-x}Ni_xFe_2O_4$ spinel nanoparticles in photodegradation and antibacterial activity. *J. Photochem. Photobiol. B Biol.* **2016**, *165*, 121–132. [CrossRef]
14. Machala, L.; Tucek, J.; Zboril, R. Polymorphous transformations of nanometric iron(III) oxide: A review. *Chem. Mater.* **2011**, *23*, 3255–3272. [CrossRef]
15. Shahmoradi, Y.; Souri, D. Growth of silver nanoparticles within the tellurovanadate amorphous matrix: Optical band gap and band tailing properties, beside the Williamson-Hall estimation of crystallite size and lattice strain. *Ceram. Int.* **2019**, *45*, 7857–7864. [CrossRef]
16. Auderbrand, N.; Auffredic, J.P.; Louer, D. X-ray diffraction study of the early stages of the growth of nanoscale zinc oxide crystallites obtained from thermal decomposition of four precursors. General concepts on precursor-dependent microstructural properties. *Chem. Mater.* **1998**, *10*, 2450–2461. [CrossRef]
17. Köseoğlu, Y. Structural, magnetic, electrical and dielectric properties of $Mn_xNi_{1-x}Fe_2O_4$ spinel nanoferrites prepared by PEG assisted hydrothermal method. *Ceram Int.* **2013**, *39*, 4221–4230. [CrossRef]
18. Minakshi, M.S.; Watcharatharapong, T.; Chakraborty, S.; Ahuja, R.; Duraisamy, S.; Rao, P.T.; Munichandraiah, N. Synthesis, and crystal and electronic structure of sodium metal phosphate for use as a hybrid capacitor in non-aqueous electrolyte. *Dalton Trans.* **2015**, *44*, 20108.
19. Minakshi, M.; Sharma, N.; Ralph, D.; Appadoo, D.; Nallathamby, K. Synthesis and characterization of $Li(Co_{0.5}Ni_{0.5})PO_4$ cathode for Li-Ion aqueous battery applications. *Electrochem. Solid State Lett.* **2011**, *14*, A86. [CrossRef]
20. Hussain, A.; Abbas, T.; Niazi, S.B. Preparation of $Ni_{1-x}Mn_xFe_2O_4$ ferrites by sol-gel method and study of their cation distribution. *Ceram Int.* **2013**, *39*, 1221–1225. [CrossRef]
21. Al-Hada, N.M.; Kamari, H.M.; Shaari, A.H.; Saion, E. Fabrication and characterization of manganese-zinc ferrite nanoparticles produced utilizing heat treatment technique. *Res. Phys.* **2019**, *12*, 1821–1825. [CrossRef]
22. Ashiq, N.M.; Ehsan, M.F.; Iqbal, M.J.; Gul, I.H. Synthesis, structural and electrical characterization of Sb^{3+} substituted spinel nickel ferrite ($NiSb_xFe_{2-x}O_4$) nanoparticles by reverse micelle technique. *J. Alloys Comp.* **2011**, *509*, 5119–5126. [CrossRef]
23. Tong, S.-K.; Chi, P.-W.; Kung, S.H.; Wei, D.H. Tuning bandgap and surface wettability of $NiFe_2O_4$ driven by phase transition. *Sci. Rep.* **2018**, *8*, 1338. [CrossRef]
24. Enuka, E.; Monne, M.A.; Lan, X.; Gambin, V.; Koltun, R.; Chen, M.Y. 3D inkjet printing of ferrite nanomaterial thin films for magneto-optical devices. In *Quantum Sensing and Nano Electronics and Photonics, Proceedings of the Photonic West 2020, San Francisco, CA, USA, 31 January 2020*; SPIE-International Society for Optics and Photonics: Bellingham, WA, USA, 2020; Volume 11288, p. 112881K.

25. Uda, M.N.A.; Gopinath, S.C.B.; Hashim, U.; Halim, N.H.; Parmin, N.A.; Afnan Uda, M.N.; Anbu, P. Production and characterization of silica nanoparticles from fly ash: Conversion of agro-waste into resource. *Prep. Biochem. Biotechnol.* **2020**, *51*, 86–95. [CrossRef]
26. Schaeffer, D.A.; Polizos, G.; Smith, D.B.; Lee, D.F.; Hunter, S.R.; Datskos, P.G. Optically transparent and environmentally durable superhydrophobic coating based on functionalized SiO_2 nanoparticles. *Nanotechnology* **2015**, *26*, 055602. [CrossRef]
27. Minakshi, M. Lithium intercalation into amorphous $FePO_4$ cathode in aqueous solutions. *Electrochim. Acta* **2010**, *55*, 9174–9178. [CrossRef]
28. Li, Q.; Kartikowati, C.W.; Horie, S.; Ogi, T.; Iwaki, T.; Okuyama, K. Correlation between particle size domain structure and magnetic properties of highly crystalline Fe_3O_4 nanoparticles. *Sci. Rep.* **2017**, *7*, 9894. [CrossRef]
29. Cullity, B.D.; Graham, C.D. *Introduction to Magnetic Materials*; Wiley: Hoboken, NJ, USA, 2011.
30. Devi, I.E.; Soibam, C. Law of approach to saturation in Mn-Zn ferrite nanoparticles. *J. Supercond. Nov. Magn.* **2019**, *32*, 1293–1298. [CrossRef]
31. Brown, W.F. Theory of the approach to magnetic saturation. *Phys. Rev.* **1940**, *58*, 736–743. [CrossRef]
32. Mondalek, F.G.; Zhang, Y.Y.; Kropp, B.; Kopke, R.D.; Ge, X.; Jackson, R.L.; Dormer, K.J. The permeability of SPION over an artificial three-layer membrane is enhanced by external magnetic field. *J. Nanobiotechnology* **2006**, *4*, 4. [CrossRef]
33. Pankhurst, Q.A.; Connolly, J.; Jones, S.K.; Dobson, J. Applications of magnetic nanoparticles in biomedicine. *J. Phys. D Appl. Phys.* **2003**, *36*, R167. [CrossRef]
34. Umut, E.; Coşkun, M.; Pineider, F.; Berti, D.; Güngüneş, H. Nickel ferrite nanoparticles for simultaneous use in magnetic resonance imaging and magnetic fluid hyperthermia. *J. Colloid Interface Sci.* **2019**, *550*, 199–209. [CrossRef]
35. Bhardwaj, A.; Parekh, K.; Jain, N. In vitro hyperthermic effect of magnetic fluid on cervical and breast cancer cells. *Sci. Rep.* **2020**, *10*, 15249. [CrossRef]
36. Huh, Y.M.; Jun, Y.W.; Song, H.T.; Kim, S.; Choi, J.S.; Lee, J.H.; Yoon, S.; Kim, K.S.; Shin, J.S.; Suh, J.S.; et al. In vivo magnetic resonance detection of cancer by using multifunctional magnetic nanocrystals. *J. Am. Chem. Soc.* **2005**, *127*, 12387. [CrossRef]
37. Chakradharya, V.K.; Ansaria, A.; Akhtara, M.J. Design, synthesis, and testing of high coercivity cobalt doped nickel ferrite nanoparticles for magnetic applications. *J. Magn. Mater.* **2019**, *469*, 674–680. [CrossRef]
38. Pardavi-Horvath, M.J. Microwave applications of soft ferrites. *J. Magn. Mater.* **2000**, *171*, 215–216. [CrossRef]
39. Sharif, S.; Yazdani, A.; Rahimi, K. Incremental substitution of Ni with Mn in $NiFe_2O_4$ to largely enhance its supercapacitance properties. *Sci. Rep.* **2020**, *10*, 10916. [CrossRef]

Article

Effect of Silica Embedding on the Structure, Morphology and Magnetic Behavior of $(Zn_{0.6}Mn_{0.4}Fe_2O_4)_\delta/(SiO_2)_{(100-\delta)}$ Nanoparticles

Thomas Dippong [1,*], **Iosif Grigore Deac** [2], **Oana Cadar** [3] and **Erika Andrea Levei** [3]

[1] Faculty of Science, Technical University of Cluj-Napoca, 76 Victoriei Street, 430122 Baia Mare, Romania

[2] Faculty of Physics, Babes-Bolyai University, 1 Kogalniceanu Street, 400084 Cluj-Napoca, Romania; iosif.deac@phys.ubbcluj.ro

[3] INCDO-INOE 2000, Research Institute for Analytical Instrumentation, 67 Donath Street, 400293 Cluj-Napoca, Romania; oana.cadar@icia.ro (O.C.); erika.levei@icia.ro (E.A.L.)

* Correspondence: dippong.thomas@yahoo.ro

Abstract: The effect of SiO_2 embedding on the obtaining of single-phase ferrites, as well as on the structure, morphology and magnetic properties of $(Zn_{0.6}Mn_{0.4}Fe_2O_4)_\delta(SiO_2)_{100-\delta}$ ($\delta = 0$–100%) nanoparticles (NPs) synthesized by sol-gel method was assessed. The phase composition and crystallite size were investigated by X-ray diffraction (XRD), the chemical transformations were monitored by Fourier transform infrared (FT-IR) spectroscopy, while the morphology of the NPs by transmission electron microscopy (TEM). The average crystallite size was 5.3–27.0 nm at 400 °C, 13.7–31.1 nm at 700 °C and 33.4–49.1 nm at 1100 °C. The evolution of the saturation magnetization, coercivity and magnetic anisotropy as a function of the crystallite sizes were studied by vibrating sample magnetometry (VSM) technique. As expected, the SiO_2 matrix shows diamagnetic behavior accompanied by the accidentally contribution of a small percent of ferromagnetic impurities. The $Zn_{0.6}Mn_{0.4}Fe_2O_4$ embedded in SiO_2 exhibits superparamagnetic-like behavior, whereas the unembedded $Zn_{0.6}Mn_{0.4}Fe_2O_4$ behaves like a high-quality ferrimagnet. The preparation route has a significant effect on the particle sizes, which strongly influences the magnetic behavior of the NPs.

Keywords: silica matrix; sol-gel synthesis; zinc-manganese ferrite; magnetic properties

Citation: Dippong, T.; Deac, I.G.; Cadar, O.; Levei, E.A. Effect of Silica Embedding on the Structure, Morphology and Magnetic Behavior of $(Zn_{0.6}Mn_{0.4}Fe_2O_4)_\delta/(SiO_2)_{(100-\delta)}$ Nanoparticles. *Nanomaterials* **2021**, *11*, 2232. https://doi.org/10.3390/nano11092232

Academic Editor: Julian Maria Gonzalez Estevez

Received: 6 August 2021
Accepted: 27 August 2021
Published: 29 August 2021

Publisher's Note: MDPI stays neutral with regard to jurisdictional claims in published maps and institutional affiliations.

1. Introduction

Zinc ferrite ($ZnFe_2O_4$) has a normal spinel structure and remarkable magnetic, electrical, electrochemical and sensing properties, making it suitable for a wide-range of applications [1–4]. Its excellent magnetization is attributed to the inversion of Fe^{3+} and Zn^{2+} ions between tetrahedral (A) and octahedral (B) sites [4]. Manganese ferrite ($MnFe_2O_4$) has a partially inverse spinel structure and numerous applications due to its tunable magnetic properties, small-sized particles, possibility to be controlled by an external magnetic field, easy synthesis process and biocompatibility. It is also an inorganic heat-resistant, non-corrosive, non-toxic and environmentally friendly material with coloristic properties [5–7].

Due to their unique properties, the Mn-Zn ferrites are widely used in many fields such as medicine, environmental depollution, energy storage, gas sensors and photocatalysis [8–10]. The use of nanoferrites in medicine is possible due to their ability to locally heat up the tissues under an external variable magnetic field as a consequence of thermal losses. The high sensitivity to oxidation and cytotoxicity of pure metal particles makes them inadequate for medical applications, while iron oxide nanoparticles are promising candidates, due to their biocompatibility, especially if they are covered with inorganic or organic biocompatible coatings (i.e., alcohol, esters, SiO_2) [11]. Recently, due to their low toxicity, Mn-Zn ferrites become a center of attraction for hyperthermia, electrical, electronic and telecommunication devices [12,13]. Cubic spinel Mn-Zn ferrites belonging to soft

71

ferrites are interesting magnetic materials due to the low core loss, corrosion resistance, high saturation magnetization (M_S), high magnetic permeability and low eddy loss [10,12].

The properties of spinel ferrites can be easily tuned and controlled as they depend on the composition and particle size distribution. The oxidation state of cations and their distribution in the spinel structure also impact the magnetic behavior of ferrites [12,14]. Mixed Mn-Zn ferrites have the spinel structure with Fe^{3+} ions occupying tetrahedral (A) and octahedral (B) sites, and the Mn^{2+} and Zn^{2+} ions occupying A sites [15]. The variation in this "normal" arrangement leads to an inverted spinel structure, with Fe^{3+} ions in A sites, and Mn^{2+} and Zn^{2+} ions in B sites. Additionally, the ferromagnetic arrangement of spinel structure with the magnetic moment in the A site will align antiparallel to any external magnetic field. Therefore, the overall magnetic behavior of Mn-Zn ferrites is associated to the particle morphology and cation distribution in the crystalline structure [8]. The structural, magnetic and electric properties may be changed by introducing various cations into the spinel structure [15].

The Mn-Zn ferrites were produced by microwave, co-precipitation, pyrolysis, decomposition, sol-gel, auto-combustion, hydrothermal and solid-state techniques. The synthesis route plays a key role in tailoring the structure, morphology and properties of the NPs [7–10]. Generally, the chemical methods give fine grained microstructure, convoyed by long reaction time and post-synthesis thermal treatment; however, often poor crystallinity and broad particle size distribution can alter the desired properties [9,12]. The sol-gel route is a common way to prepare ferrite NPs due to its simplicity, low cost and good control over the structure and properties. Encapsulating ferrite NPs into solid silica (SiO_2) allows the control of the particle growth, minimization of particle agglomeration, as well as the enhancement of their magnetic guidability and biocompatibility [5].

This paper aims to investigate the structure, morphology and magnetic properties of $(Zn_{0.6}Mn_{0.4}Fe_2O_4)_\delta(SiO_2)_{100-\delta}$ NPs produced by sol-gel route as well as the effect of various factors such as the ferrite type, embedding material, synthesis method, stoichiometric composition and calcination temperature.

2. Materials and Methods

All chemical reagents, of analytical grade, were purchase from Merck (Darmstadt, Germany) and used without further purification. $(Zn_{0.6}Mn_{0.4}Fe_2O_4)_\delta(SiO_2)_{100-\delta}$ ($\delta = 0–100\%$) NPs were synthesized by sol-gel method. Zinc nitrate ($Zn(NO_3)_2 \cdot 6H_2O$), manganese nitrate ($Mn(NO_3)_2 \cdot 3H_2O$) and ferric nitrate ($Fe(NO_3)_3 \cdot 9H_2O$) were dissolved in 1,4-butanediol (BD) in a molar ratio of 0.6:0.4:2:8. To the nitrate-BD mixture, an ethanolic solution of tetraethyl orthosilicate (TEOS) acidified with nitric acid (pH = 2) in NO_3^-:TEOS molar ratio of 0:2 ($\delta = 0\%$), 0.5:1.5 ($\delta = 25\%$), 1:1 ($\delta = 50\%$), 1.5:0.5 ($\delta = 75\%$) and 2:0 ($\delta = 100\%$) was added dropwise with continuous stirring, at room temperature. The obtained mixture was exposed to open air for gelation (8 weeks). The glassy gels were grinded and dried at 300 °C for 4 h, then calcined in air, at 400, 700 and 1100 °C for 5 h in a LT9 muffle furnace (Nabertherm, Lilienthal, Germany).

The morphology of NPs was investigated on dried suspensions of NPs onto carbon-coated copper grids using a Hitachi HD-2700 (Hitachi, Tokyo, Japan) transmission electron microscope (TEM) equipped with digital image recording system. The effect of the ferrite content embedded in the amorphous SiO_2 matrix on structural properties was investigated by X-ray diffraction pattern recorded at room temperature, using a D8 Advance (Bruker, Karlsruhe, Germany) diffractometer, operating at 40 kV and 40 mA with CuKα radiation ($\lambda = 1.54060$ Å). The formation of the ferrite and SiO_2 matrix was monitored using a Spectrum BX II (Perkin Elmer, Waltham, MA, USA) Fourier-transform infrared spectrometer on KBr pellets containing 1% sample. A cryogen free VSM magnetometer (Cryogenic Ltd., London, UK) was used for magnetic measurements. The saturation magnetization was measured in high magnetic field up to 10 T, while the magnetic hysteresis loops were performed between −2 to 2 T, at 300 K on samples incorporated in an epoxy resin. The particles were not aligned in an applied magnetic field.

3. Results and Discussion

Figure 1 shows the distribution, shape and size of $(Zn_{0.6}Mn_{0.4}Fe_2O_4)_\delta(SiO_2)_{100-\delta}$ NPs (δ = 25–100%) calcined at 1100 °C. The TEM images of SiO_2 (δ = 0%) does not allow the identification of the SiO_2 network. In case of samples calcined at low temperatures the TEM images are blurry with particles that are not clearly separated. The TEM images show irregular, spongy aggregates of spherical, large (50 nm for δ = 100%, 41 nm for δ = 75%) NPs in case of high ferrite content and small (34 nm for δ = 25%, 35 nm for δ = 50%) NPs in case of high SiO_2 content. The different size and morphology of NPs could be the consequence of the different kinetics of metal oxides formation reaction and of the different particle growth rate following volume expansion and supersaturation reduction [13–15].

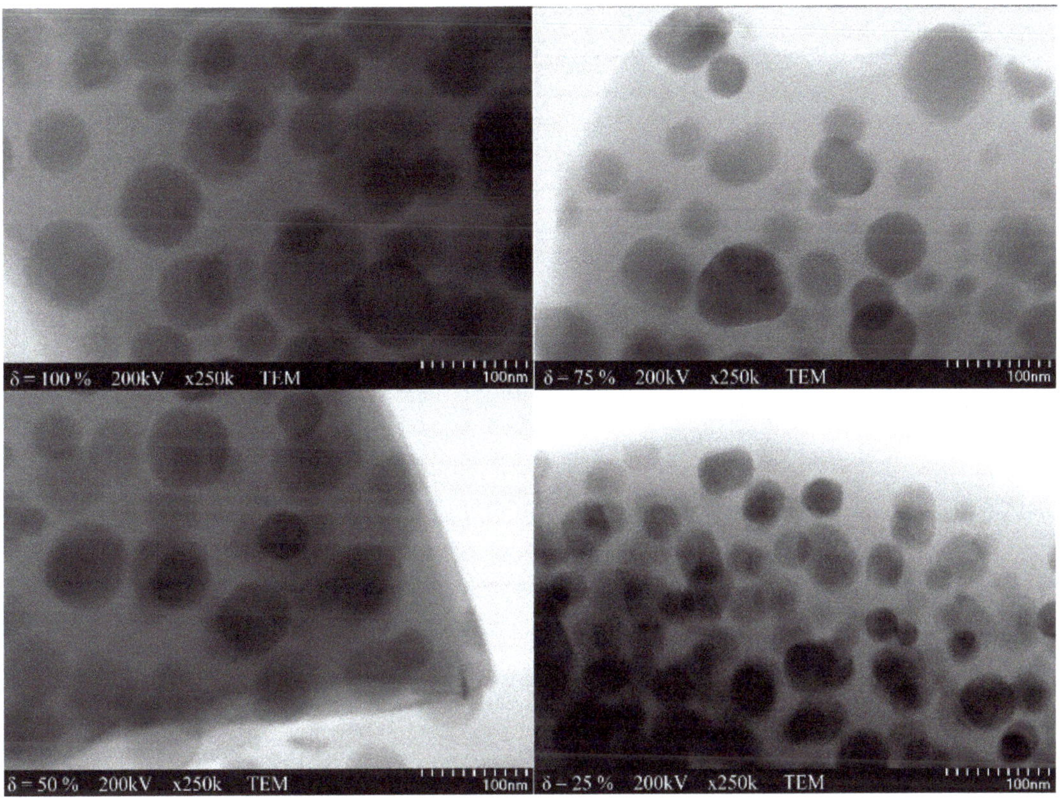

Figure 1. TEM images of $(Zn_{0.6}Mn_{0.4}Fe_2O_4)_\delta(SiO_2)_{100-\delta}$ NPs calcined at 1100 °C.

Figure 2 shows the XRD patterns and FT-IR spectra of NPs calcined at 400, 700 and 1100 °C. The degree of crystallinity was estimated as the ratio between the area of all diffraction peaks and the total area of diffraction peaks and amorphous halo [13,16]. In case of δ = 0%, the formation of amorphous SiO_2 matrix is confirmed by the broad halo in the 2θ range of 15–30°, at all temperatures. At low calcination temperatures (400 and 700 °C), crystalline $Zn_{0.6}Mn_{0.4}Fe_2O_4$ ($ZnFe_2O_4$ (JCPDS card 16-6205 [17]) and $MnFe_2O_4$ (JCPDS card 10-0319 [17]) phases with cubic spinel structure belonging to $Fd3m$ group are remarked. At 400 °C, in case of the NPs with δ = 25% and 50%, the amorphous halo in the 2θ range of 15–30° is also remarked, but much flatten than in case of δ = 0%.

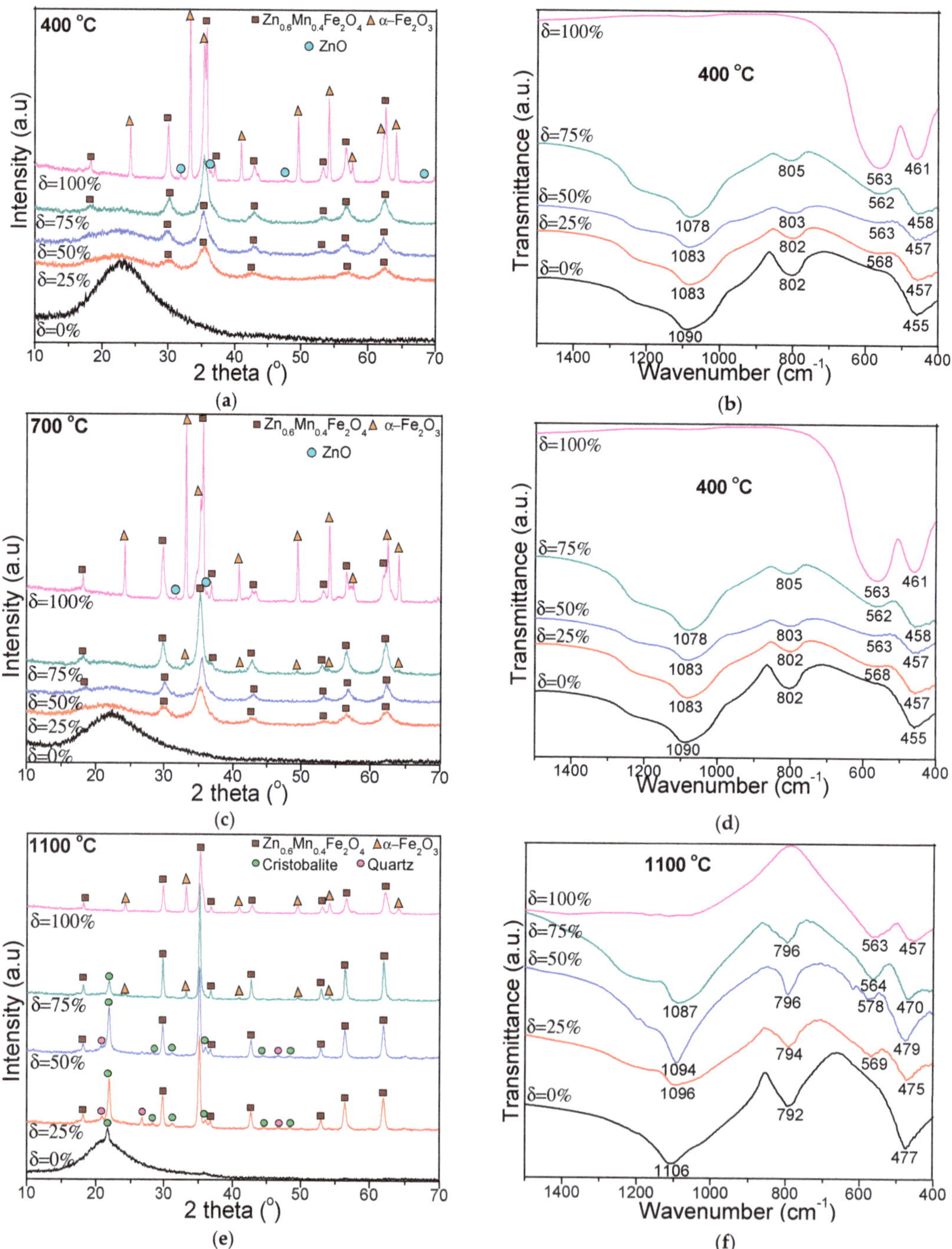

Figure 2. XRD patterns (**a**,**c**,**e**) and FT-IR spectra (**b**,**d**,**f**) of $(Zn_{0.6}Mn_{0.4}Fe_2O_4)_\delta(SiO_2)_{100-\delta}$ NPs calcined at 400, 700 and 1100 °C.

At 700 °C, at higher ferrite content ($\delta = 75\%$), the main $Zn_{0.6}Mn_{0.4}Fe_2O_4$ phase is accompanied by $\alpha\text{-}Fe_2O_3$ secondary phase (JCPDS card 87-1164) [17]) most probably due to the fact that ferrite is only partially embedded in the SiO_2 matrix, as a consequence of the low SiO_2 content. The presence of $\alpha\text{-}Fe_2O_3$ might be also attributed to the insufficient calcination temperature or time needed to obtain pure crystalline $Zn_{0.6}Mn_{0.4}Fe_2O_4$ phase [18]. In case of $Zn_{0.6}Mn_{0.4}Fe_2O_4$ ($\delta = 100\%$), well-crystallized $Zn_{0.6}Mn_{0.4}Fe_2O_4$ and $\alpha\text{-}Fe_2O_3$ phases together with low-crystallized ZnO phase (JCPDS card 36-1451) [17]) is observed. Consequently, it is possible that the SiO_2 matrix may favor the formation of single-crystalline $Zn_{0.6}Mn_{0.4}Fe_2O_4$ phase.

At 1100 °C, in case of $\delta = 0\%$, the broad halo belonging to the SiO_2 matrix displays a sharp peak attributed to cristobalite (JCPDS card 89-3434 [17]), which leads to the assumption that the SiO_2 matrix is partially crystalline, in the absence of ferrite. In case of NP with $\delta = 25\%$ and 50%, beside the main $Zn_{0.6}Mn_{0.4}Fe_2O_4$ phase, the secondary phases belonging to the SiO_2 matrix (cristobalite and quartz, JCPDS card 71-0785 [17]) is also remarked. At higher ferrite content ($\delta = 75\%$), the crystalline quartz phase disappears, the cristobalite content is reduced and the main $Zn_{0.6}Mn_{0.4}Fe_2O_4$ phase is accompanied by the low-crystallized $\alpha\text{-}Fe_2O_3$ phase. In case of $Zn_{0.6}Mn_{0.4}Fe_2O_4$ ($\delta = 100\%$) calcined at 1100 °C, the degree of crystallinity of $\alpha\text{-}Fe_2O_3$ decreases, as suggested the less intense diffraction peaks compared to the NPs calcined at 400 and 700 °C. The peaks corresponding to ferrite embedded in SiO_2 matrix, become more intense at 1100 °C, indicating high degree of crystallinity, high crystallite size (because coalescence process) and high nucleation rate (due to the small growth rate and homogenously distributed nanoparticles) and low effect of inert surface layer of the crystals [13,19]. Furthermore, the position of the highest peak is shifted towards higher angles with increasing $Zn_{0.6}Mn_{0.4}Fe_2O_4$ content in the SiO_2 amorphous matrix.

The average crystallite size calculated using the Debye-Scherrer's equation [13,16] considering the highest intensity peak (311) and the quantitative phase analysis using reference intensity ratio method of the NPs are presented in Table 1. The crystalline $SiO_{2\%}$ represents the sum of SiO_2-cristobalite % and SiO_2-quartz %. The content of crystalline SiO_2 decreases, while that of $\alpha\text{-}Fe_2O_3$ increases with the increase of ferrite content embedded in the SiO_2 matrix. In addition, the average crystallites size increases from 5.3 to 49.1 nm with the increase of calcination temperature and ferrite content embedded in SiO_2 [20–22]. The peak intensity also increases with the calcination temperature, leading to enhanced crystallinity of the $Zn_{0.6}Mn_{0.4}Fe_2O_4$. In addition, the crystalline volume to surface ratio is increasing alongside calcination temperature, as showed in the TEM data, due to particle size expansion [21].

Table 1. The crystallite size and quantitative analysis according to XRD of $(Mn_{0.6}Mn_{0.4}Fe_2O_4)_\delta$ $(SiO_2)_{100-\delta}$ NPs calcined at 400, 700 and 1100 °C.

Sample	Temperature (°C)	D (nm)	Quantitative Analysis (%)
	400	5.3	100% $Zn_{0.6}Mn_{0.4}Fe_2O_4$
$\delta = 25\%$	700	13.7	100% $Zn_{0.6}Mn_{0.4}Fe_2O_4$
	1100	33.4	52% $Zn_{0.6}Mn_{0.4}Fe_2O_4$/48% SiO_2
	400	6.7	100% $Zn_{0.6}Mn_{0.4}Fe_2O_4$
$\delta = 50\%$	700	20.5	100% $Zn_{0.6}Mn_{0.4}Fe_2O_4$
	1100	34.5	59% $Zn_{0.6}Mn_{0.4}Fe_2O_4$/41% SiO_2
	400	7.8	100% $Zn_{0.6}Mn_{0.4}Fe_2O_4$
$\delta = 75\%$	700	22.6	89% $Zn_{0.6}Mn_{0.4}Fe_2O_4$/11% $\alpha\text{-}Fe_2O_3$
	1100	39.8	71% $Zn_{0.6}Mn_{0.4}Fe_2O_4$/10% $\alpha\text{-}Fe_2O_3$/19% SiO_2
	400	27.0	46% $Zn_{0.6}Mn_{0.4}Fe_2O_4$/44% $\alpha\text{-}Fe_2O_3$/10% ZnO
$\delta = 100\%$	700	31.1	49% $Zn_{0.6}Mn_{0.4}Fe_2O_4$/45% $\alpha\text{-}Fe_2O_3$/6% ZnO
	1100	49.1	73% $Zn_{0.6}Mn_{0.4}Fe_2O_4$/27% $\alpha\text{-}Fe_2O_3$

At all temperatures, the FT-IR spectra of NPs with δ = 25–100% show the absorption bands of Zn-O and Mn-O bonds stretching vibration at 562–578 cm^{-1} and of Fe–O bonds vibration at 457–479 cm^{-1} [20]. The intensity of the vibration band at 562–578 cm^{-1} increases with the increase of calcination temperature, most probably due to the increase of the ferrite's crystallinity, as ferrites acts as continuously bonded crystals with atoms linked to all nearest neighbors by equivalent ionic, covalent or van der Waals forces [21,23,24]. The increase of calcination temperature determines a small shift of the vibration band due to the changes in the ion's distribution between A and B sites [23–26]. The specific bands of the SiO$_2$ matrix were identified in the FT-IR spectra of NPs with δ = 0–75%: 1078–1106 cm^{-1} with a shoulder around 1200 cm^{-1} assigned to vibration of Si–O–Si chains, 792–809 cm^{-1} assigned to the vibrations of SiO$_4$ tetrahedron and 457–479 cm^{-1} assigned to the Si-O bond vibration that is overlapping the band of Fe–O vibration [13,19]. The low polycondensation degree of the SiO$_2$ network is suggested by the high intensity of these bands.

The SiO$_2$ matrix (δ = 0%) calcined at 700 and 1100 °C displays diamagnetic behavior at high magnetic fields, while at low magnetic fields it shows the signature of the presence of a low concentration of some ferromagnetic impurities (Figure 3).

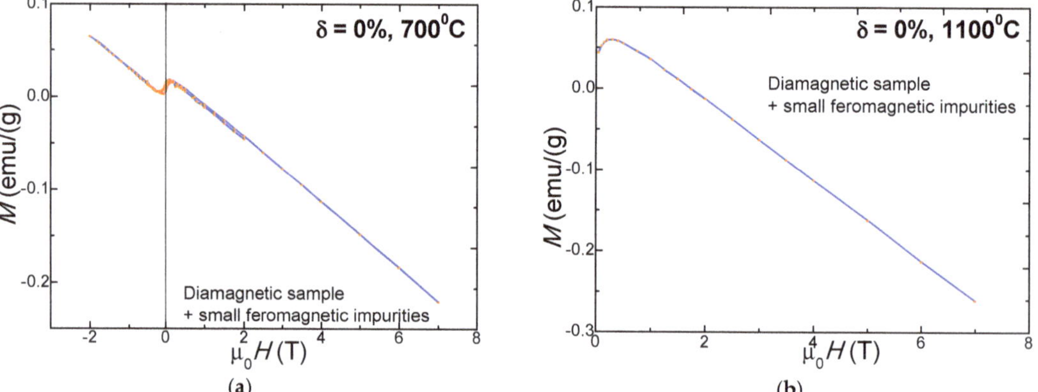

Figure 3. Diamagnetic properties of the SiO$_2$ matrix (δ = 0%) for the samples calcined at 700 °C (**a**) and 1100 °C (**b**).

The room temperature magnetic hysteresis loops, the saturation magnetization (M_S), remanent magnetization (M_R), coercive field (H_c) values of $(Zn_{0.6}Mn_{0.4}Fe_2O_4)_\delta(SiO_2)_{100-\delta}$ (δ = 25–100%) NPs calcined at 700 and 1100 °C are shown in Figure 4.

Typical 'S' shape hysteresis loops are obtained for all the investigated NPs, indicating their soft magnetic behavior. For the NPs calcined at low temperature (400 °C), the magnetic parameters values are very small, making difficult to quantify the effect of the heat treatment on the sample magnetic properties. Contrarily, the hysteresis loops show saturation under the same magnetic field for the NPs calcined at 700 °C and 1100 °C. The calcination temperature greatly influences the morphology and the phase constitution of the samples, and this has an important effect on the magnetic properties of the prepared compounds [26]. The hysteresis loops indicate low coercivity (Hc) values suggesting that the coalescence of the crystallites results in enhanced magnetic coupling and higher magnetization [16]. The H_C of the spinel nano-ferrites is mainly dictated by the magnetocrystalline anisotropy, particle-particle interaction, strain, morphology and the grain sizes [27]. The M_S, as well as the magnetic anisotropy of the NPs calcined at 700 and 1100 °C, increase with increasing of the $Zn_{0.6}Mn_{0.4}Fe_2O_4$ content in the SiO$_2$ matrix.

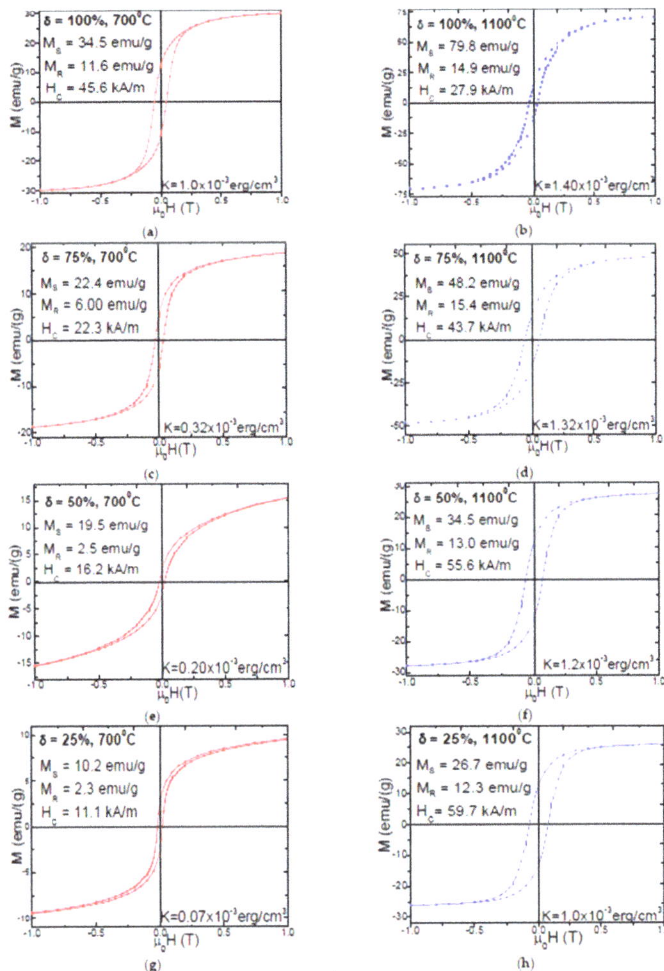

Figure 4. Magnetic hysteresis loops of $(Zn_{0.6}Mn_{0.4}Fe_2O_4)_\delta(SiO_2)_{100-\delta}$ ($\delta = 100\%$ (**a,b**), 75% (**c,d**), 50% (**e,f**) and 25% (**g,h**)) NPs calcined at 700 and 1100 °C.

With increasing calcination temperature, the NPs have a softer magnetic behavior since higher magnetic field are needed to fully saturate the magnetization of the samples [28]. Generally, the magnetic parameters (M_S and M_R) of the NPs improve with increasing calcination temperature since the crystallinity of $Zn_{0.6}Mn_{0.4}Fe_2O_4$ phase increases, the interatomic distances are larger, more vacancies could appear and the coordination number could diminish [13]. Besides the dilution of the magnetic matrix, the SiO_2 matrix generates surface disorder effect on the particles, it can increase the number of broken chemical bond, leading to spin canting, pinning of the magnetic field and change of the particle size distribution [13]. As known, the cation distribution in the spinel lattice has a strong influence on the magnitude of the saturation magnetization [13,28].

The NPs calcined at 700 °C have low M_S values due to their low crystallinity degree, high defect concentration, low coordination number and large interatomic spacing [13,19]. The local defects lead to the weakening of the superexchange interaction between the A and B sites [13]. The bulk density of the samples, as well as the grain sizes also can have an important effect on the magnetic behavior of the samples. The release of tension by the bigger grains is higher than by the smaller grains leading to lattice expansion [29]. The pores can act as pinning centers for the magnetic moments and for the domain walls [12]. The H_C increases for the NPs calcined at 700 °C and decreases for the NPs calcined at 1100 °C, with increasing the ferrite content in the SiO_2 matrix. The low Hc, values for the NPs with $\delta = 25, 50\%$ calcined at 700 °C, indicating that these samples can be easily demagnetized in technical applications [12]. The higher H_C values are related with Zn atoms, while the presence of Mn atoms in the ferrite results in a decrease of the coercive field [16,27,28]. The variation of the coercivity H_c with increasing Zn-Mn ferrite content in the SiO_2 matrix is related with the change of the anisotropy constant, particle size distribution and with the magnetic domains structure of the NPs [30]. The above mentioned, superparamagnetic-like behavior of these NPs arises from the small crystallite sizes with low anisotropy which are easily thermal activated [13,16,19]. As known, in polycrystalline ferrites, the variation of the main magnetic parameters (i.e., M_S, M_R and H_C) are related with the grain sizes, bulk density, magnetic anisotropy, superexchange interaction between A and B sites and the surface effects [12].

The magnetization of $MnFe_2O_4$ is mainly determined by the competition between A-B super exchange and B-B exchange [31]. When Zn^{2+} is added in the Mn ferrite structure, the A-B exchange in weakened and B–B sublattice interaction becomes stronger. In case of NPs ($\delta = 75$ and 100%), the presence of the α-Fe_2O_3 as secondary phase contributes to the overall magnetization. The M_S increases with increasing content of ferrite and α-Fe_2O_3 phases. The decrease of magnetization can be the consequence of the spin canting in the B sublattice as described by the non-collinear Yafet-Kittel model [18,31–33]. At the particles surface a dead layer is formed which contains broken chemical bonds, deviations from the bulk cation distribution, non-saturation effects, randomly oriented magnetic moments, lattice defects, etc. which usually result in the depreciation of the magnetic properties of the NPs [34]. The large grains contain a great number of magnetic domains walls and the motions of these walls will have a dominant contribution to the magnetization process versus the rotation of the magnetic moments [13,16,19]. As the particle size decreases multiple domains are converted into single larger domains and align along the direction of applied field. As the motion of domain walls is responsible for reversing the field, the decrease of particle size may transform a magnetic material from multi domain phase to a single domain phase resulting in the H_c increase [18].

The anisotropy constant K was calculated using Equation (1) [35]:

$$K = \frac{\mu_0 \cdot M_S \cdot H_c}{2} \tag{1}$$

where M_S is the saturation magnetization, μ_0 is the magnetic permeability of the free space ($\mu_0 = 1.256 \times 10^{-6}$ N/A^2) and Hc is the coercivity field (T).

The linear variation of the M_S as a function of the crystallite size for the NPs calcined at 700 and 1100 °C is presented in Figure 5. A non-linear increase of the coercivity (H_C) (NPs calcined at 700 °C) and magnetic anisotropy constant (K) (NPs calcined at 700 and 1000 °C) and decrease of the H_C (NPs calcined at 1100 °C), with the increase of crystallite size is observed. The behavior of H_C suggests single magnetic domains for the NPs calcined at 700 °C as H_C increases with increasing average particle size and multidomain regime for the NPs calcined at 1100 °C as the H_C decreases with increasing average particle size. The reduction of the H_C with increasing ferrite content can be connected with the particle sizes.

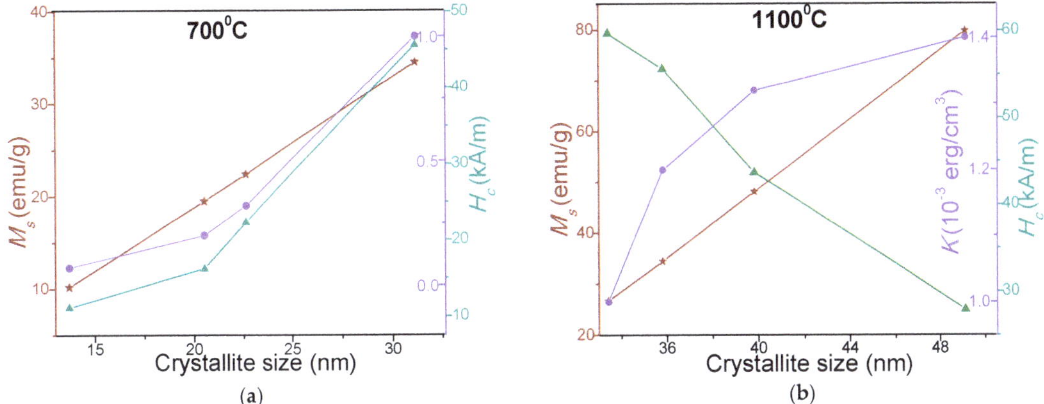

Figure 5. Variation of M_S, H_C and K with crystallite size of NPs calcined at 700 °C (**a**) and 1100 °C (**b**).

The TEM and XRD data indicates an increase of the particle and crystallites sizes with increasing ferrite content embedded in the SiO_2 matrix. Similarly, an increase of the K with increasing ferrite content embedded in the SiO_2 matrix was found. Therefore, we concluded that the increase of the particle sizes leads to the enhancement of the K. When the size of a magnetic particle increases above a critical crystallite diameter, a multidomain region occurs, where the H_C reduces its value [36]. The SiO_2 matrix generates stress on the surface of the ferrite particle, which will hamper the rotation of the magnetic moments from the dead layer at the surface contributing in this way to the reduction of the coercivity [13]. The magnetic anisotropy constant depends on the lattice crystalline symmetry, the crystalline anisotropy, and on particles size and shape [13,27]. The magnetocrystalline anisotropy is also affected by the distribution of the magnetic ions on the surface of the nanosized particles. In our case, the Mn^{2+} and Zn^{2+} ions have considerable contributions to the magnetocrystalline anisotropy. The magnetocrystalline energy enhances with increasing particle sizes and the volumes of NPs [12]. The highest K value is obtained for the $Zn_{0.6}Mn_{0.4}Fe_2O_4$ NPs (δ = 100%) for which a high magnetic field is necessary to saturate the magnetization, due to the magnetic ions disorder generated by the surface effects in the dead layer [13].

The Mn-Zn ferrite embedding in SiO_2 matrix allows an easy control of the crystallization temperature, nanoparticle sizes and magnetic properties of the NPs. The easily tunable magnetic and electrical properties of the NPs recommend them for application as ferrofluids, hybrid supercapacitors, biocompatible magnetic-fluids, medical applications or as efficient magnetically recyclable material for the removal of chemical and biological contaminants from industrial wastewaters [37–43].

4. Conclusions

$(Zn_{0.6}Mn_{0.4}Fe_2O_4)_\delta(SiO_2)_{100-\delta}$ (δ = 25–100%) NPs with different morphologies, phase constitutions and magnetic properties were obtained using sol-gel method. The stoichiometric composition, synthesis technique and particle size play a critical role in defining the ferrite properties. At low calcination temperatures (400 and 700 °C), single crystalline phase was obtained, excepting NC with δ = 75% at 700 °C, where the main phase was accompanied by the secondary α-Fe_2O_3 phase. At high calcination temperatures (1100 °C), cristobalite and quartz phases were also present. The average crystallites size increases with increasing calcination temperature, as well as with increasing ferrite content embedded in the SiO_2 matrix: 16.15 ± 10.85 nm (400 °C), 22.4 ± 8.7 nm (700 °C) și 41.25 ± 7.85 nm (1100 °C). The TEM images show irregular aggregates of spherical, large NPs for samples with high ferrite content (50 nm) or small NPs for samples with high SiO_2 content (34 nm) NPs. The saturation magnetization (M_S) and anisotropy constant (K) of the NPs calcined at 700 °C (10.2–34.5 emu/g and $0.07 \cdot 10^{-3}$–$1.00 \cdot 10^{-3}$ erg/cm^3) and at 1100 °C (26.7–79.8 emu/g and $1.00 \cdot 10^{-3}$–$1.4 \cdot 10^{-3}$ erg/cm^3) increase with increasing ferrite content embedded in the SiO$_2$ matrix. The coercivity field (H_C) increases (11.1–45.6 kA/m) for the NPs calcined at 700 °C and decreases (59.7–27.9 kA/m) for the NPs calcined at 1100 °C with increasing ferrite content embedded in the SiO$_2$ matrix. It was found a linear dependency of the M_S on the crystallite's sizes for the both calcination temperatures (700 and 1000 °C).

Author Contributions: T.D. conceptualization, investigation, writing, supervision and editing the work, I.G.D., O.C. and E.A.L. methodology, formal analysis and writing the manuscript. All authors have read and agreed to the published version of the manuscript.

Funding: The APC was funded by Technical University of Cluj-Napoca, Grant Support GA11, GA12, GA25/24 September 2019.

Acknowledgments: This research was funded by the Romanian Ministry of Research and Innovation through CCCDI-UEFISCDI, grant for the project PN- III-P4-ID-PCCF-2016-0112. The authors would like to express their gratitude to Lucian Barbu-Tudoran for the TEM measurements.

Conflicts of Interest: The authors declare no conflict of interest. The funders had no role in the design of the study; in the collection, analyses, or interpretation of data; in the writing of the manuscript, or in the decision to publish the results.

References

1. Chand, P.; Vaish, S.; Kumar, P. Structural, optical and dielectric properties of transition metal (MFe$_2$O$_4$; M = Co, Ni and Zn) nanoferrites. *Phys. B Condens. Matter* **2017**, *524*, 53–63. [CrossRef]
2. Peng, S.; Wang, S.; Liu, R.; Wu, J. Controlled oxygen vacancies of ZnFe$_2$O$_4$ with superior gas sensing properties prepared via a facile one-step self-catalyzed treatment. *Sens. Actuators B Chem.* **2019**, *288*, 649–655. [CrossRef]
3. Sarkar, K.; Mondal, R.; Dey, S.; Kumar, S. Cation vacancy and magnetic properties of ZnFe$_2$O$_4$ microspheres. *Phys. B Condens. Matter* **2020**, *583*, 412015. [CrossRef]
4. Ge, Y.C.; Wang, Z.L.; Yi, M.Z.; Ran, L.P. Fabrication and magnetic transformation from paramagnetic to ferrimagnetic of ZnFe$_2$O$_4$ hollow spheres. *Trans. Nonferrous Met. Soc. China* **2019**, *29*, 1503–1509. [CrossRef]
5. Asghar, K.; Qasim, M.; Das, D. Preparation and characterization of mesoporous magnetic MnFe$_2$O$_4$@mSiO$_2$ nanocomposite for drug delivery application. *Mater. Today Proc.* **2020**, *26*, 87–93. [CrossRef]
6. Junlabhut, P.; Nuthongkum, P.; Pechrapa, W. Influences of calcination temperature on structural properties of MnFe$_2$O$_4$ nanopowders synthesized by co-precipitation method for reusable absorbent materials. *Mater. Today Proc.* **2018**, *5*, 13857–13864. [CrossRef]
7. Sivakumar, A.; Dhas, S.S.J.; Dhas, S.A.M.B. Assessment of crystallographic and magnetic phase stabilities on MnFe$_2$O$_4$ nano crystalline materials at shocked conditions. *Solid State Sci.* **2020**, *107*, 106340. [CrossRef]
8. Martins, M.L.; Florentino, A.O.; Cavalheiro, A.A.; Silva, R.I.V.; Dos Santos, D.I.; Saeki, M.J. Mechanisms of phase formation along the synthesis of Mn–Zn ferrites by the polymeric precursor method. *Ceram. Int.* **2014**, *40*, 16023–16031. [CrossRef]
9. Naik, P.P.; Hasolkar, S.S.; Keluskar, S.; Pisseurlekar, V. Role of Mn^{+2} ions in monitoring structural, optical, magnetic and electrical properties of manganese zinc ferrite nanoparticles. *J. Mater. Sci. Mater. Electron.* **2021**, in press. [CrossRef]
10. Kaewmanee, T.; Wannapop, S.; Phuruangrat, A.; Thongtem, T.; Wiranwetchayan, O.; Promnopas, W.; Sansongsiri, S.; Thongtem, S. Effect of oleic acid content on manganese-zinc ferrite properties. *Inorg. Chem. Comm.* **2019**, *103*, 87–92. [CrossRef]
11. Balanov, V.A.; Kiseleva, A.P.; Krivoshapkina, E.F.; Kashtanov, E.A.; Gimaev, R.R.; Zverev, V.I.; Krivoshapkin, P.V. Synthesis of (Mn$_{(1-x)}$Zn$_x$)Fe$_2$O$_4$ nanoparticles for magnetocaloric applications. *J. Sol-Gel Sci. Technol.* **2020**, *95*, 795–800. [CrossRef]
12. Anwar, H.; Maqsood, A.; Gul, I.H. Effect of synthesis on structural and magnetic properties of cobalt doped Mn–Zn nano ferrites. *J. Alloy. Compd.* **2018**, *626*, 410–414. [CrossRef]
13. Dippong, T.; Cadar, O.; Deac, I.G.; Lazar, M.; Borodi, G.; Levei, E.A. Influence of ferrite to silica ratio and thermal treatment on porosity, surface, microstructure and magnetic properties of Zn$_{0.5}$Ni$_{0.5}$Fe$_2$O$_4$/SiO$_2$ nanocomposites. *J. Alloy. Compd.* **2020**, *828*, 154409. [CrossRef]
14. Aparna, M.L.; Nirmala Grace, A.; Sathyanarayanan, P.; Sahu, N.K. A comparative study on the supercapacitive behaviour of solvothermally prepared metal ferrite (MFe$_2$O$_4$, M=Fe, Co, Ni, Mn, Cu, Zn) nanoassemblies. *J. Alloy. Compd.* **2018**, *745*, 385–395. [CrossRef]
15. Arteaga-Cardona, F.; Pal, U.; Alonso, J.M.; de la Presa, P.; Mendoza-Alvarez, M.E.; Salazar-Kuri, U.; Mendez-Rojas, M.A. Tuning magnetic and structural properties of MnFe$_2$O$_4$ nanostructures by systematic introduction of transition metal ions M^{2+} (M = Zn, Fe, Ni, Co). *J. Magn. Magn. Mater.* **2019**, *490*, 165496. [CrossRef]
16. Dippong, T.; Levei, E.A.; Deac, I.G.; Neag, E.; Cadar, O. Influence of Cu^{2+}, Ni^{2+}, and Zn^{2+} ions doping on the structure, morphology, and magnetic properties of Co-ferrite embedded in SiO$_2$ matrix obtained by an innovative sol-gel route. *Nanomaterials* **2020**, *10*, 580. [CrossRef]

17. *Powder Diffraction File, Joint Committee on Powder Diffraction Standards*; International Center for Diffraction Data: Swarthmore, PA, USA, 1999.

18. Zawar, S.; Atiq, S.; Riaz, S.; Naseem, S. Correlation between particle size and magnetic characteristics of Mn-substituted $ZnFe_2O_4$ ferrites. *Superlattices Microstr.* **2016**, *93*, 50–56. [CrossRef]

19. Dippong, T.; Levei, E.A.; Deac, I.G.; Goga, F.; Cadar, O. Investigation of structural and magnetic properties of $Ni_xZn_{1-x}Fe_2O_4/SiO_2$ ($0 \leq x \leq 1$) spinel-based nanocomposites. *J. Anal. Appl. Pyrol.* **2019**, *144*, 104713. [CrossRef]

20. Dippong, T.; Levei, E.A.; Cadar, O.; Mesaros, A.; Borodi, G. Sol-gel synthesis of $CoFe_2O_4$: SiO_2 nanocomposites–insights into the thermal decomposition process of precursors. *J. Anal. Appl. Pyrol.* **2017**, *125*, 169–177. [CrossRef]

21. Al-Hada, N.M.; Kamari, H.M.; Shaari, A.H.; Saion, E. Fabrication and characterization of manganese–zinc ferrite nanoparticles produced utilizing heat treatment technique. *Res. Phys.* **2019**, *12*, 1821–1825. [CrossRef]

22. Hu, P.; Yang, H.; Pan, D.; Wang, H.; Tian, J.J.; Zhang, S.; Wang, X.; Volinsky, A.A. Heat treatment effects on microstructure and magnetic properties of Mn–Zn ferrite powders. *J. Magn Magn Mater.* **2010**, *322*, 173–177. [CrossRef]

23. Salunkhe, A.B.; Khot, V.M.; Phadatare, M.R.; Thorat, N.D.; Joshi, R.S.; Yadav, H.M.; Pawar, S.H. Low temperature combustion synthesis and magnetostructural properties of Co-Mn nanoferrites. *J. Magn. Magn. Mater.* **2014**, *352*, 91–98. [CrossRef]

24. Reddy, M.P.; Zhou, X.; Yann, A.; Du, S.; Huang, Q.; Mohamed, A. Low temperature hydrothermal synthesis, structural investigation and functional properties of $Co_xMn_{1-x}Fe_2O_4$ ($0 \leq x \leq 1$) nanoferrites. *Superlattices Microst.* **2015**, *81*, 233–242. [CrossRef]

25. Kotsikau, D.; Ivanovskaya, M.; Pankov, V.; Fedotova, Y. Structure and magnetic properties of manganese zinc-ferrites prepared by spray pyrolysis method. *Solid State Sci.* **2015**, *39*, 69–73. [CrossRef]

26. Mali, A.; Ataie, A. Structural characterization of nano-crystalline $BaFe_{12}O_{19}$ powders synthesized by sol–gel combustion route. *Scr. Mater.* **2005**, *53*, 1065–1070. [CrossRef]

27. Abdallah, H.M.I.; Moyo, T.; Msomi, J.Z. The effect of annealing temperature on the magnetic properties of $Mn_xCo_{1-x}Fe_2O_4$ ferrites nanoparticles. *J. Supercond. Nov. Magn.* **2012**, *25*, 2625–2630. [CrossRef]

28. Atif, M.; Sato Turtelli, R.; Grössinger, R.; Siddique, M.; Nadeem, M. Effect of Mn substitution on the cation distribution and temperature dependence of magnetic anisotropy constant in $Co_{1-x}Mn_xFe_2O_4$ ($0.0 \leq x \leq 0.4$) ferrites. *Ceram. Int.* **2014**, *40*, 471–748. [CrossRef]

29. Deepty, M.; Srinivas, C.; Ranjith Kumar, E.; Krisha Mohan, N.; Prajapat, C.L.; Chandrasekhar Rao, T.V.; Singh Meena, S.; Kumar Verma, A.; Sastry, D.L. XRD, EDX, FTIR and ESR spectroscopic studies of co-precipitated Mn–substituted Zn–ferrite nanoparticles. *Ceram. Int.* **2019**, *45*, 8037–8044. [CrossRef]

30. Hou, X.; Feng, J.; Liu, X.; Ren, Y.; Fan, Z.; Zhang, M. Magnetic and high rate adsorption properties of porous $Mn_{1-x}Zn_xFe_2O_4$ ($0 < x < 0.8$) adsorbents. *J. Colloid Interface Sci.* **2011**, *353*, 524–529.

31. Wiriya, N.; Bootchanont, A.; Maensiri, S.; Swatsitang, E. Magnetic properties of $Zn_{1-x}Mn_xFe_2O_4$ nanoparticles prepared by hydrothermal method. *Microelectron. Eng.* **2014**, *126*, 1–8. [CrossRef]

32. Verma, R.; Chauhan, A.; Batoo, K.M.; Hadi, M.; Raslan, E.H.; Kumar, R.; Ijaz, M.F.; Assaifan, A.K. Structural, optical, and electrical properties of vanadium-doped, lead- free BCZT ceramics. *J. Alloy. Compd.* **2021**, *869*, 159520. [CrossRef]

33. Nasrin, S.; Chowdhury, F.U.Z.; Hoque, S.M. Study of hydrodynamic size distribution and hyperthermia temperature of chitosan encapsulated zinc-substituted manganese nano ferrites suspension. *Phys. B Condens. Matter* **2019**, *561*, 54–63. [CrossRef]

34. Arulmurugan, R.; Vaidyanathan, G.; Sendhilnathan, S.; Jeyadevan, B. Mn-Zn ferrite nanoparticles for ferrofluid preparation: Study on thermal-magnetic properties. *J. Magn. Magn. Mater.* **2006**, *298*, 83–94. [CrossRef]

35. Atif, M.; Asghar, M.W.; Nadeem, M.; Khalid, W.; Ali, Z.; Badshah, S. Synthesis and investigation of structural, magnetic and dielectric properties of zinc substituted cobalt ferrites. *J. Phys. Chem. Solids* **2018**, *123*, 36–42. [CrossRef]

36. Cullity, B.D.; Graham, C.D. *Introduction to Magnetic Materials*; John Wiley & Sons, Inc.: Hoboken, NJ, USA, 2009; p. 359.

37. Rocher, V.; Siaugue, J.-M.; Cabuil, V.; Bee, A. Removal of organic dyes by magnetic alginate beads. *Water Res.* **2008**, *42*, 1290–1298. [CrossRef]

38. Zhao, X.; Shi, Y.; Cai, Y.; Mou, S. Cetyltrimethylammonium bromide-coated magnetic nanoparticles for the preconcentration of phenolic compounds from environmental water samples. *Environ. Sci. Technol.* **2008**, *42*, 1201–1206. [CrossRef]

39. Unruh, K.M.; Chien, C.L. *Nanomaterials: Synthesis, Properties and Applications*; Edelstein, A.S., Cammrats, R.C., Eds.; Institute of Physics Publishing: Bristol, UK, 1996; p. 447.

40. Wang, Y.; Gu, H. Core-shell-type magnetic mesoporous silica nanocomposites for bioimaging and therapeutic agent delivery. *Adv. Mater.* **2015**, *27*, 576–585. [CrossRef]

41. Pollert, E.; Veverka, P.; Veverka, M.; Kaman, O.; Závěta, K.; Vasseur, S.; Epherre, R.; Goglio, G.; Duguet, E. Search of new core materials for magnetic fluid hyperthermia: Preliminary chemical and physical issues. *Prog. Solid State Chem.* **2009**, *37*, 1–14. [CrossRef]

42. Hankiewicz, J.H.; Alghamdil, N.; Hammelev, N.M.; Anderson, N.R.; Campley, R.E.; Stupic, K.; Przybylski, M.; Zukrowski, J.; Celinski, Z.J. Zinc doped copper ferrite particles as temperature sensors for magnetic resonance imaging. *AIP Adv.* **2017**, *7*, 056703. [CrossRef]

43. Jang, J.-T.; Nah, H.; Lee, J.-H.; Moon, S.H.; Kim, M.G.; Cheon, J. Critical enhancements of MRI contrast and hyperthermic effects by dopant-controlled magnetic nanoparticles. *Angew. Chem. Int. Edit.* **2009**, *48*, 1234–1238. [CrossRef] [PubMed]

nanomaterials

Article

CuBi$_2$O$_4$ Synthesis, Characterization, and Application in Sensitive Amperometric/Voltammetric Detection of Amoxicillin in Aqueous Solutions

Raluca Dumitru (m.Vodă) [1], Sorina Negrea [2,3], Cornelia Păcurariu [1], Adrian Surdu [4], Adelina Ianculescu [4], Aniela Pop [1] and Florica Manea [1,*]

[1] Faculty of Industrial Chemistry and Environmental Engineering, Politehnica University Timisoara, Piata Victoriei No. 2, 300006 Timisoara, Romania; raluca.voda@upt.ro (R.D.(m.V.)); cornelia.pacurariu@upt.ro (C.P.); aniela.pop@upt.ro (A.P.)

[2] National Institute of Research and Development for Industrial Ecology (INCD ECOIND), 300431 Timisoara, Romania; negrea.sorina@yahoo.com

[3] Department of Environmental Engineering and Management, "Gheorghe Asachi" Technical University of Iasi, 700050 Iasi, Romania

[4] Department of Oxide Materials Science and Engineering, Faculty of Applied Chemistry and Materials Science, Polytehnic University of Bucharest, Gh. Polizu Street No. 1–7, 011061 Bucharest, Romania; adrian.surdu@live.com (A.S.); a_ianculescu@yahoo.com (A.I.)

* Correspondence: florica.manea@upt.ro; Tel.: +40-256-403-070

Citation: Dumitru (m.Vodă), R.; Negrea, S.; Păcurariu, C.; Surdu, A.; Ianculescu, A.; Pop, A.; Manea, F. CuBi$_2$O$_4$ Synthesis, Characterization, and Application in Sensitive Amperometric/Voltammetric Detection of Amoxicillin in Aqueous Solutions. *Nanomaterials* **2021**, *11*, 740. https://doi.org/10.3390/nano11030740

Academic Editor: Doong-Joo Kim

Received: 16 February 2021
Accepted: 12 March 2021
Published: 15 March 2021

Publisher's Note: MDPI stays neutral with regard to jurisdictional claims in published maps and institutional affiliations.

Abstract: CuBi$_2$O$_4$ synthesized by thermolysis of a new Bi(III)-Cu(II) oxalate coordination compound, namely Bi$_2$Cu(C$_2$O$_4$)$_4$·0.25H$_2$O, was tested through its integration within carbon nanofiber paste electrode, namely CuBi/carbon nanofiber (CNF), for the electrochemical detection of amoxicillin (AMX) in the aqueous solution. Thermal analysis and IR spectroscopy were used to characterize a CuBi$_2$O$_4$ precursor to optimize the synthesis conditions. The copper bismuth oxide obtained after a heating treatment of the precursor at 700 °C/1 h was investigated by an X-ray diffraction and scanning electron microscopy. The electrochemical behavior of CuBi/CNF in comparison with CNF paste electrode showed the electrocatalytic activity of CuBi$_2$O$_4$ toward amoxicillin detection. Two potential detections, with one at the potential value of +0.540 V/saturated calomel electrode (SCE) and the other at the potential value of −1.000 V/SCE, were identified by cyclic voltammetry, which were exploited to develop the enhanced voltammetric and/or amperometric detection protocols. Better electroanalytical performance for AMX detection was achieved for CuBi/CNF using differential-pulsed and square-wave voltammetries than others reported in the literature. Very nice results obtained through anodic and cathodic currents recorded at +0.750 V/SCE and −1.000 V/SCE in the same time period using a pseudo multiple-pulsed amperometry technique showed the great potential of the CuBi/CNF paste electrode for practical applications in amoxicillin detection in aqueous solutions.

Keywords: CuBi$_2$O$_4$ electrocatalyst; amoxicillin; electrochemical detection; CuBi$_2$O$_4$-carbon nanofiber paste electrode

1. Introduction

Currently, the sensing field is continuously evolving to reach demand in various practical applications, e.g., medical, pharmaceuticals, food, and environmental quality. Electrochemical sensors represent the main category of sensors due to their advantages of fast response, simplicity, and versatility. However, their electroanalytical performance depends on the electrode type correlated with the electrochemical techniques. Carbon-based electrodes are intensively used in the electroanalysis, but the main disadvantage is given by the slow rate of the electron transfer at the carbon surface that confers low sensitivity and implicit, limited applications. Nanostructured carbon offers a partial solution for the mentioned problem due to its enhanced electrocatalytic activity depending on the carbon type, but its high price restricts the practical applications. The carbon nanofiber is one of the cheapest categories of nanostructured carbon, which is vastly studied in electroanalysis both as a working electrode [1,2] and as the substrate for further modification to enhance its sensing characteristics [3–5]. Copper bismuth oxide ($CuBi_2O_4$) is receiving growing attention as a promising material for application in photocatalysis [6,7], photoelectrochemical water splitting [8,9], and sensing [10–13]. The properties of $CuBi_2O_4$ are in relation to the synthesis method. To date, many methods have been developed for preparing $CuBi_2O_4$ as coprecipitation [14], solvothermal [15] and hydrothermal methods [16–18], combustion [19], sono-chemical reactions [20], etc.

In this study, the electrocatalytic properties of copper bismuthate was studied using carbon nanofiber (CNF) as a substrate and for a comparison in detecting amoxicillin in the aqueous solution. Amoxicillin (AMX) belongs to the third generation of antibiotic-penicillin class and very frequently prescribed against a wide range of infections. It is also used in human and veterinary medicine. In this context, its presence in food (e.g., milk, eggs, meat) has been reported [21] and, also, it was detected in water environments in drinking water via hospital effluents and municipal wastewater as main sources [22]. AMX belongs to the emerging pollutants class from antibiotics-based pharmaceuticals being one of the eight substances listed in the "watch list" of substances for union-wide monitoring in the field of a water policy pursuant to Directive 2008/105/EC of the European Parliament and of the Council and repealing Commission Implementing Decision (EU) 2015/495 [23]. As other pharmaceuticals, its quantitative and qualitative detection methods are urgently required and several methods including chromatography, spectroscopy, and spectro-fluorometry have been used to determine amoxicillin [24–26]. Due to drawbacks related to cost, time consumption, a separate stage of sample preparation, and research for the development of a fast and easy determination method of AMX is required. The electrochemical methods should be regarded as a viable alternative for AMX determination by taking into account their advantages of a fast and simple detection method. Several electrochemical detection methods of AMX using carbon-based and modified electrodes have been reported [26–28], but more improvement of the electroanalytical performance is required for which new composition of the electrode, which represents the core of the electrochemical detection, should be developed and tested. Copper bismuth oxide synthesized by thermolysis of Bi(III)-Cu(II) oxalate coordination compound at 700 °C/h was used to modify the carbon nanofiber paste by simple mixing in paraffin oil as a CuBi/CNF electrode to enhance the performance of the electrochemical detection of amoxicillin. The new electrode was tested comparatively with CNF paste electrode in AMX detection using conventional and advanced voltammetric and amperometric techniques.

2. Materials and Methods

2.1. Synthesis and Characterization of $CuBi_2O_4$

For the synthesis of the oxalate coordination compound, $Bi(NO_3)_3 \cdot 5H_2O$, $Cu(NO_3)_2 \cdot 3H_2O$, 1,2-ethanediol, and 2 M nitric acid solution were used as reagents from Merck (Darmstadt, Germany). Copper bismuth oxide was obtained through thermal decomposition of the precursor in the temperature range of 500–700 °C. A water solution containing bismuth nitrate, copper nitrate, 1,2-ethanediol, and nitric acid (2 M) in a molar ratio 2:1:4:2.66 was used. This mixture was heated in a water bath for about 30 min. The reaction is finished when no more gas is evolved. The resulting solid product was purified by washing with acetone and dried under a room temperature environment. The coordination compound $Bi_2Cu(C_2O_4)_4 \cdot 0.25H_2O$ is synthesized using 2 M nitric acid solution.

In order to obtain copper bismuth oxide powders, the oxalate coordination compound was thermally treated in air, for 1 h, with a heating rate of 10 °C min^{-1} in the temperature range of 500–700 °C.

FTIR (Fourier-transform infrared) spectrum (KBr pellets) of the coordination compound was recorded on a Jasco FT-IR spectrophotometer (Jasco, Tokyo, Japan), in the range of 4000–400 cm^{-1}. Thermal measurements through thermal gravimetry (TG), differential thermogravimetry (DTG) and differential scanning calorimetry (DSC) were performed on the precursor using a NETZSCH-STA 449C instrument (Netzsch Group, Selb, Germany) in the range of 25–700 °C, using alumina crucibles. The experiments were carried out in artificial air flow of 20 mL min^{-1} and a heating rate of 10 K min^{-1}.

The phase purity and crystal structure of calcined powders were determined by using X-ray diffraction (XRD) analyses performed at room temperature by means of a Rigaku Ultima IV diffractometer (Rigaku Co., Tokyo, Japan), using Ni-filtered CuKα radiation (λ = 1.5418 Å), with a scan step increment of 0.01° and a scanning rate of 1 °/min, for 2θ ranged between 20–80°.

The size and the agglomeration tendency of the CuBi$_2$O$_4$ particles was assessed by a scanning electron microscopy (FE-SEM), using a high resolution FEI QUANTA INSPECT F microscope with a field emission gun (FEI Co., Eindhoven, The Nederlands).

2.2. Copper Bismuthate-Carbon Nanofiber Paste Electrode (CuBi/CNF) Obtaining and Electrochemical Characterization

The copper bismuthate-carbon nanofiber paste electrode (CuBi/CNF) was obtained by mixing certain amounts of carbon nanofibers, paraffin oil, and copper bismuth to reach the ratio of 21.5 wt. % carbon nanofibers, 43 wt. % copper bismuth, and 35.5 wt. % paraffin oil. For comparison, a carbon nanofiber paste electrode (CNF) was similarly obtained with the composition of 64.5 wt. % carbon nanofibers and 35.5 wt. % paraffin oil. The mass ratio of copper bismuth, carbon nanofibers, and paraffin oil of 2:1:1.65 was chosen to assure the sufficient contribution of copper bismuthate, the electrode stability, and conductivity. For comparison, the ratio of 3:1.65 carbon nanofibers to paraffin oil as the carbon nanofiber paste was used. The carbon nanofibers (>98% purity) and paraffin oil were of an analytical standard, provided by Sigma Aldrich (Darmstadt, Germany).

Prior to each detection experiment, the electrode was electrochemically activated and stabilized by 9 cyclic voltammograms using a cyclic voltammetry (CV) technique within the potential range between −1.5 and +1.0 V/saturated calomel electrode (SCE in the 0.1 M Na$_2$SO$_4$ supporting electrolyte.

2.3. Electrochemical Detection of AMX

All the electrochemical experiments were performed using a classical three electrodes cell, having the saturated calomel electrode (SCE) as a reference electrode, platinum as a counter electrode, and the CuBi/CNF and respective CNF paste electrodes as the working electrode. The electrodes were connected to an Autolab potentiostat/galvanostat PGSTAT 302 (Eco Chemie, The Netherlands) controlled with GPES 4.9 software.

The experiments were performed in the 0.1 M Na$_2$SO$_4$ 0.1 M supporting electrolyte, and the applied techniques were cyclic voltammetry (CV), differential-pulsed voltammetry (DPV), and square-wave voltammetry (SWV) as voltammetric techniques as well as chronoamperometry (CA) and multiple-pulsed amperometry (MPA) as amperometric techniques.

The lowest limit of detection (*LOD*) and limit of quantification (*LOQ*) were calculated through the following equation, i.e., LOD = 3·SD·m^{-1} and LOQ = 10·SD·m^{-1}, where *SD* is the standard deviation of 5 blanks and *m* is the slope of the analytical plots [29].

3. Results

3.1. Characterization of Bi$_2$Cu(C$_2$O$_4$)$_4$·0.25H$_2$O Oxalate Precursor

The synthesis method of the oxalate coordination compound is based on the redox reaction between 1,2-ethanediol and nitrate ion:

$$12\,C_2H_4(OH)_2 + 6\,([Bi(OH_2)_6]^{3+} + 3NO) + 3\,([Cu(OH_2)_4]^{2+} + 2NO) + 8\,(H^+ + NO) \xrightarrow{\Delta t^{\circ}} 3 \quad (1)$$
$$Bi_2Cu(C_2O_4)_4 \cdot 0.25H_2O_{(s)} + 32NO_{(g)} + 87.25H_2O_{(g)}$$

The IR spectrum of the synthesized coordination compound is provided in Figure 1.

$$NO_{(g)} + \frac{1}{2} O_{2(g)} \rightarrow NO_{2(g)} \tag{2}$$

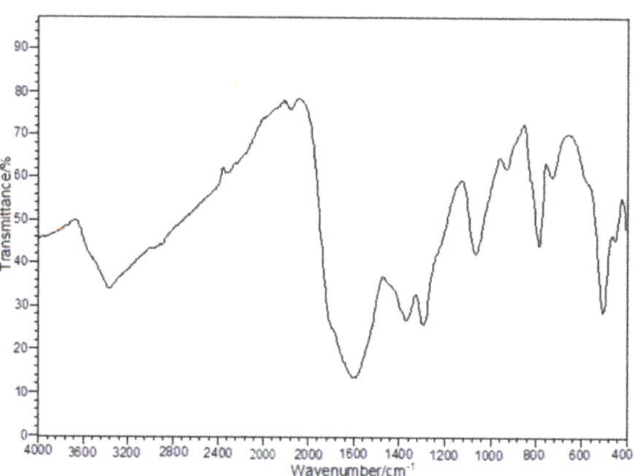

Figure 1. The IR vibrational spectrum of $Bi_2Cu(C_2O_4)_4 \cdot 0.25H_2O$ compound.

Besides the water presence [3371 cm^{-1} (ν_{OH}, ν_{H2O}), 700–800 cm^{-1} (lattice water)], two different coordination modes for oxalate anions as tetradentate bridges [1602 cm^{-1} ($\nu_{asym\,OCO}$), 1368 cm^{-1} ($\nu_{sym\,OCO}$) and 930 cm^{-1} (δ_{OCO})] and as chelate bidentate [1709 cm^{-1} ($\nu_{asym\,OCO}$), 1294 cm^{-1} ($\nu_{sym\,OCO}$)] were identified [30,31]. The band at 1064 cm^{-1} is assigned to the vibration $\nu_{(C-O)}$. The coordination of oxalate anion is confirmed by the bands lying in 500–400 cm^{-1} [$\nu_{(Bi-O)}$ and $\nu_{(Cu-O)}$ vibrations] [32,33].

The thermal decomposition of the investigated coordination compound occurs in the temperature range of 25–700 °C (Figure 2) and confirm the formation as the end decomposition product of a compound with a molecular formula of $CuBi_2O_4$ (mass loss calcd./found (%): 34.91/34.00). The first decomposition stage of the $Bi_2Cu(C_2O_4)_4 \cdot 0.25H_2O$ compound associated with an endothermic effect is attributed to the dehydration reaction that implies the evolution of the lattice water molecules (25–150 °C, mass loss: found 0.41%; calcd. 0.54%). The second decomposition step (150–400 °C) associated with an exothermic effect is assigned to the degradation of the oxalate anions (mass loss, found 33.59%, calcd. 34.37%) with the formation of amorphous $CuBi_2O_4$.

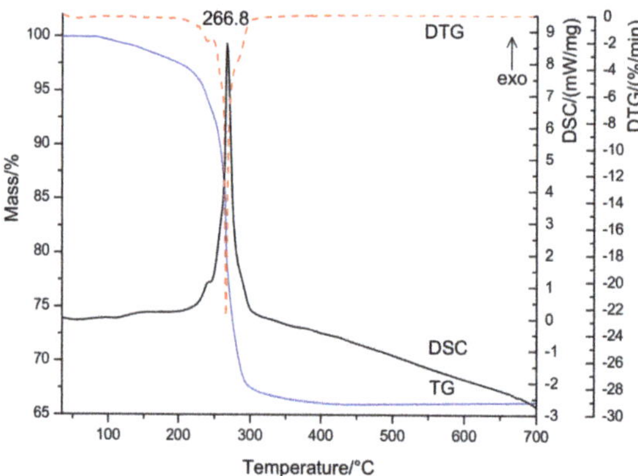

Figure 2. Thermal curves DSC, TG, and DTG of $Bi_2Cu(C_2O_4)_4 \cdot 0.25H_2O$ compound.

The crystallized copper bismuth oxide was obtained starting with 500 °C. Based on the above presented results, this study proposed the following mechanism for the thermal decomposition of a bismuth-copper oxalate precursor.

$$[Bi_2Cu(C_2O_4)_4] \cdot 0.25H_2O_{(s)} \xrightarrow{(I)} 0.25H_2O_{(g)} + [Bi_2Cu(C_2O_4)_4]_{(s)} \tag{3}$$

$$[Bi_2Cu(C_2O_4)_4]_{(s)} \xrightarrow{(II)} 4\,CO_{2(g)} + 4\,CO_{(g)} + CuBi_2O_{4(s)} \tag{4}$$

3.2. Characterization of the CuBi$_2$O$_4$ Powder

The X-ray diffraction patterns prove that the crystallization process starts at 500 °C, when CuBi$_2$O$_4$ with a tetragonal structure was identified as a major phase by its main diffraction lines. Small amounts of Bi$_2$O$_3$ was also detected as a secondary phase in the powders calcined at 500 and 600 °C, respectively (Figure 3). A heating treatment of the coordination compound as a precursor performed at 700 °C for one hour determines the formation of pure CuBi$_2$O$_4$, as shown by the XRD pattern (Figure 3). The structural parameters of the powders under investigation obtained by Rietveld refinement are summarized in Table 1. As expected, the increase of the calcination temperature induces the increase of the average crystallite size and, consequently, a decrease of the internal strains. One can observe that the increase of the crystallite size involves a decreasing evolution of the "in-plane" *a* and *b* parameters of the tetragonal unit cell of CuBi$_2$O$_4$, while the variation of the out-of-plane parameter *c* is non-monotonic.

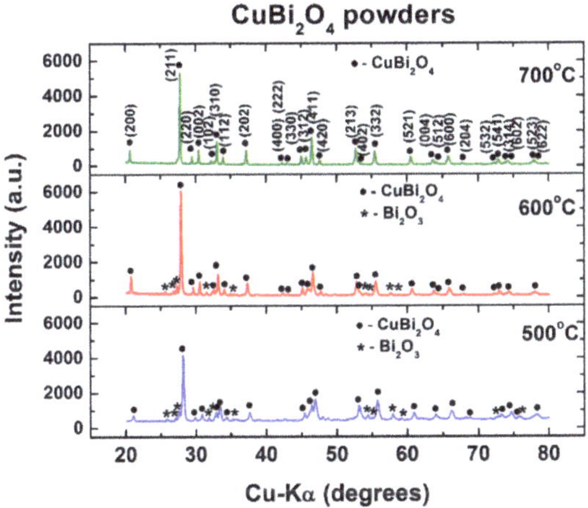

Figure 3. X-ray diffraction (XRD) patterns at room temperature for the CuBi$_2$O$_4$ powders calcined at various temperatures.

FE-SEM investigations were performed only on the single phase CuBi$_2$O$_4$ powder calcined at 700 °C for 1 h. The low magnification image of Figure 4a shows the presence of irregular particles, which exhibit a clear tendency to form aggregates due to a presintering process induced by the higher calcination temperature [34]. The high magnification image of Figure 4b reveals that most of the particles exhibit a polyhedral shape with well-defined faces, edges, and corners and various sizes that ranged between 200 and 1800 nm. An average particle size of 893.3 nm was estimated based on the histogram of the particle size distribution of Figure 4c. Taking into account the value of the average crystallite size determined from the XRD data, one can assume that polycrystalline particles, consisting of a variable number of crystallites, were obtained after calcination at 700 °C for 1 h.

Table 1. Phase composition and structural parameters of $CuBi_2O_4$ powders prepared by the oxalate route and thermally treated for 1 h at various temperatures.

Calcination Temperature (°C)		500	600	700
Phase composition		• $CuBi_2O_4$ (ICDD no. 01-080-1906)—61%, • Bi_2O_3-m (ICDD no. 04-007-1342)—27.8%, • Bi_2O_3-t (ICDD no. 01-073-6885)—11.2%.	• $CuBi_2O_4$ (ICDD no. 01-080-1906)—83.2%, • Bi_2O_3-m (ICDD no. 04-007-1342)—6.9%, • Bi_2O_3-t (ICDD no. 01-073-6885)—9.9%.	• $CuBi_2O_4$ (ICDD no.01-080-1906)—100%.
$CuBi_2O_4$ structure		Tetragonal, P4/ncc	Tetragonal, P4/ncc	Tetragonal, P4/ncc
Unit cell parameters	a (Å)	8.501143 ± 0.000912	8.500031 ± 0.000566	8.496553 ± 0.000555
	b (Å)	8.501143 ± 0.000912	8.500031 ± 0.000566	8.496553 ± 0.000555
	c (Å)	5.817769 ± 0.000817	5.827083 ± 0.000496	5.822091 ± 0.000478
	$\alpha = \beta = \gamma$ (°)	90	90	90
Unit cell volume, V (Å³)		420.4469	421.0097	420.3050
Expected R, R_{exp}		11.26198	10.68391	10.97338
R profile, R_p		9.08664	6.68487	7.20859
Weighted R profile, R_{wp}		13.50014	9.18731	9.57705
Goodness of fit, χ^2		1.43697	0.73946	0.7617
Crystallite size, <D> (nm)		29.79 ± 8.24	33.64 ± 5.65	54.00 ± 4.87
Internal strains, <S> (%)		0.29 ± 0.04	0.26 ± 0.06	0.17 ± 0.08

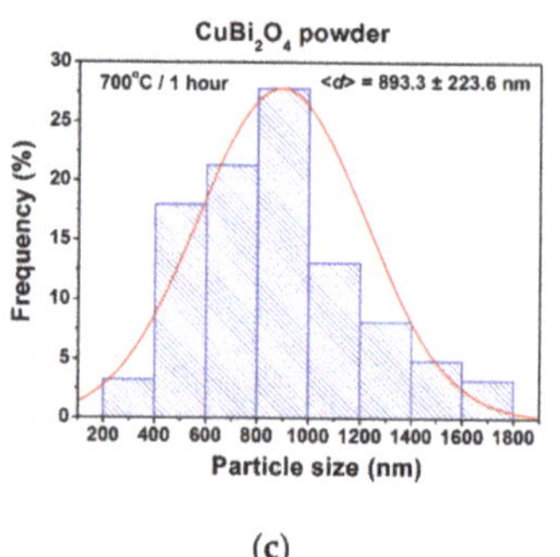

Figure 4. FE-SEM images the $CuBi_2O_4$ powder calcined at 700 °C for 1 h: (**a**) low magnification ($\times5000$) overall view, (**b**) high magnification view ($\times50,000$), and (**c**) histogram indicating the particle size distribution.

3.3. Application in Electrochemical Sensing of Amoxicillin (AMX)

The electrocatalytic activity of $CuBi_2O_4$ was tested by modifying a carbon nanofiber paste electrode as CuBi/CNF to enhance the electrocatalytic detection of amoxicillin (AMX).

Before detection testing, the CuBi/CNF electrode was characterized electrochemically by cyclic voltammetry (CV) using classical potassium ferri/ferrocyanide redox system to determine its electroactive area. Cyclic voltammograms (CV) of 4 mM $K_3[Fe(CN)_6]$ in a 1 M KNO_3 supporting electrolyte were recorded at different scanning rates. The results are presented in Figure 5a,b.

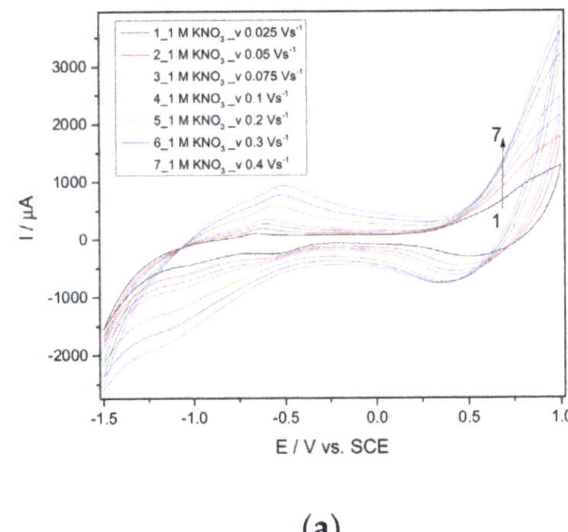
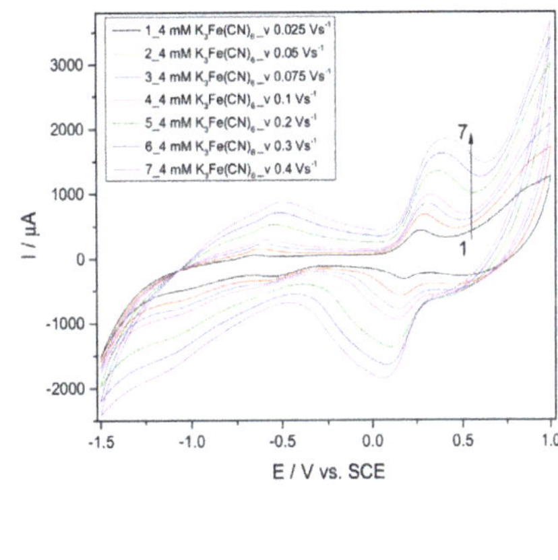

Figure 5. Cyclic voltammograms (CV) recorded at a CuBi/CNF modified electrode in a 1 M KNO$_3$ supporting electrolyte (**a**) and the presence of 4 mM K$_3$[Fe(CN)$_6$] (**b**) at different scan rates: 0.025, 0.05, 0.075, 0.1, 0.2, 0.3, and 0.4 Vs^{-1}.

The diffusion coefficient was determined comparatively for both CuBi/CNF and CNF paste electrodes according to the Randles–Sevcik Equation (5):

$$I_p = 2.69 \times 10^5 A D^{1/2} n^{3/2} v^{1/2} C \tag{5}$$

where A represents the area of the electrode (cm^2), n is the number of electrons participating in the reaction (and is equal to 1), D is the diffusion coefficient of the molecule in the solution, C is the concentration of the probe molecule in the solution and is 4 mM, and v is the scan rate (V s^{-1}). The linear dependence between the current peak and the square root of the scan rate allowed determining the diffusion coefficient of 2.70×10^{-5} cm^2 s^{-1} for the CuBi/CNF electrode and respective 7.86×10^{-6} cm^2 s^{-1} for the CNF electrode. Taking into account the theoretical diffusion coefficient value of 6.70×10^{-6} cm^2 s^{-1} found in the literature data [35], the value of the electroactive electrode area was determined to be 0.790 cm^2 for the CuBi/CNF electrode of 0.230 cm^2 for the CNF electrode vs. a 0.196-cm^2 geometric area value of the electrode. It is clear that CuBi/CNF exhibited a much higher electroactive area in comparison with one of the CNF electrodes.

3.3.1. Cyclic Voltammetry

Cyclic voltammetry (CV) was applied to characterize the electrochemical behavior of CuBi/CNF electrode in comparison with a CNF paste electrode in a 0.1-M Na$_2$SO$_4$ supporting electrolyte and in the presence of different AMX concentrations.

The presence of CuBi$_2$O$_4$ on the electrode surface determined a significant increase of the background current due to its capacitive behavior related to the morphostructural properties and the electro active surface area. It can be noticed that the oxidation of AMX started at the potential value of +0.5 V/SCE for both electrodes and the anodic peak current increased linearly with AMX concentration (see Figure 6). The linearity between the anodic current recorded at the potential value of +0.54 V/SCE and the AMX concentration allowed us to determine the detection sensitivity, which is higher for modified CuBi/CNF (181 μA mM^{-1} cm^{-2}) in comparison with the CNF paste electrode (133 μA mM^{-1} cm^{-2}) (the results of linearizations are not shown here). The electrochemical response is based on the oxidation peak, which characterizes the one electron involving an oxidation reaction of the phenolic substituent to a respective carbonyl group on the side chain of the AMX molecule [28]. One corresponding cathodic peak due to the reduction process is noticed for CuBi/CNF in comparison with the CNF electrode that did not exhibit the cathodic peak in this anodic range. This aspect suggests that the cathodic peak corresponded to the $CuBi_2O_4$ presence. According to the literature [10], in the anodic potential range, Cu^{3+} species generated during scanning in the $CuBi_2O_4$ matrix are involved in the AMX oxidation and, by reverse scanning, the reduction of Cu^{3+} to Cu^{2+} species can be noticed. Thus, a quasi-reversible redox couple at the potential value of about +0.54 V/SCE resulted from the intrinsic redox properties of $CuBi_2O_4$. However, a reduction peak that increased with AMX concentration rising appeared at the potential value of about −1.0 V/SCE, which was noticed for the CuBi/CNF electrode (Figure 5). According to the literature [14], the reduction peak corresponds to the BiO_2^- reducing to BiO_2^{2-}, which is involved in the reduction process of AMX or the AMX oxidation product.

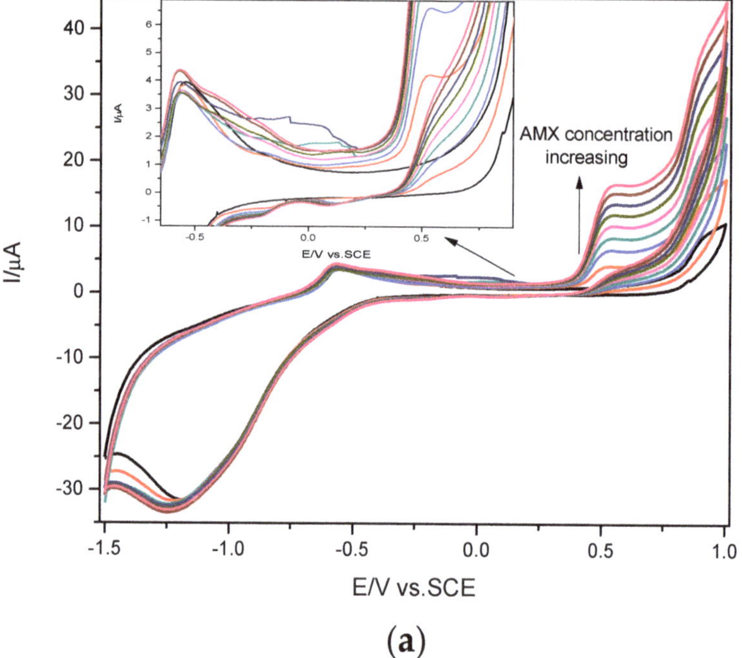

(a)

Figure 6. *Cont.*

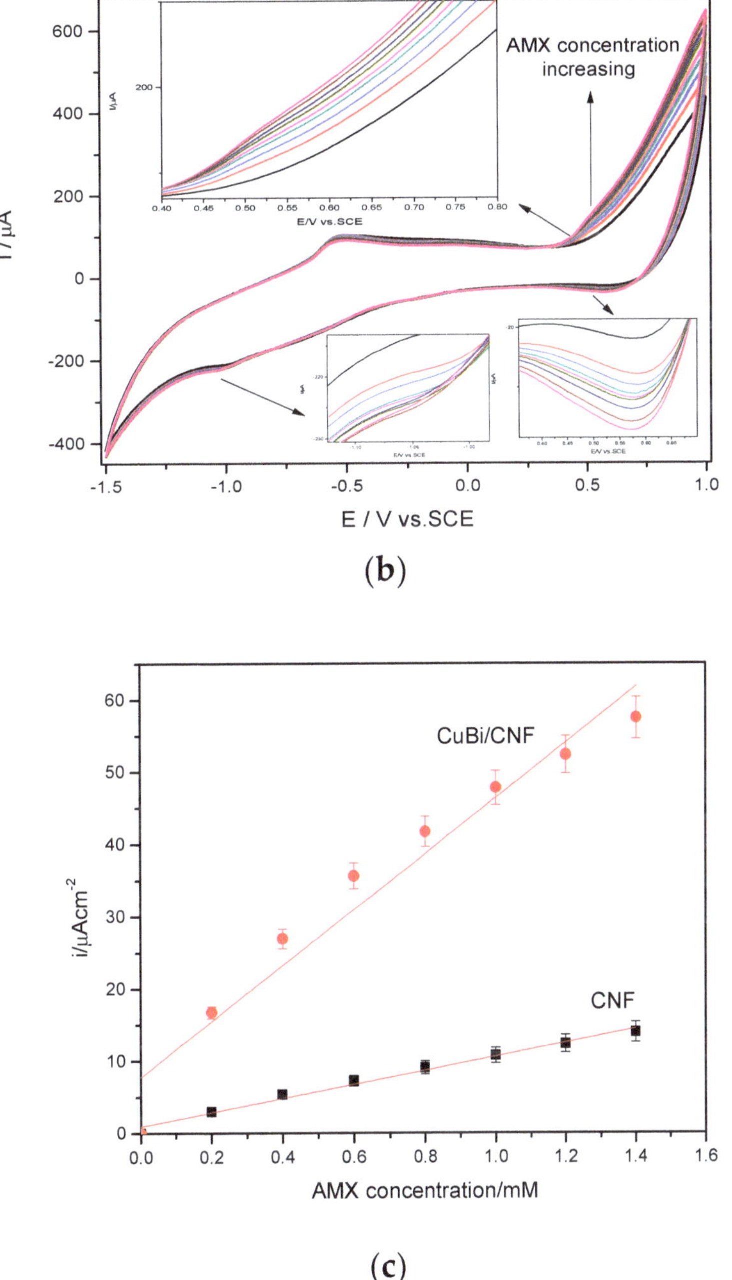

(b)

(c)

Figure 6. Cyclic voltammograms recorded at the 0.5 Vs^{-1} scan rate on the CNF paste electrode (**a**) and CuBi/CNF modified paste electrode (**b**) in 0.1 M Na$_2$SO$_4$ supporting electrolyte and in the presence of various AMX concentrations ranged from 0.2 mM to 1.6 mM AMX concentration. Comparative calibration plots for AMX detection in the concentration range from 0.2 to 1.4 mM, rerecorded at the potential value of +0.55 V/SCE (**c**).

3.3.2. Influence of the Scan Rate

To elucidate some mechanistic aspects, the influence of the scan rate on the electrochemical behaviour of CuBi/CNF in the absence/presence of 1 mM AMX and the results are presented in Figure 7.

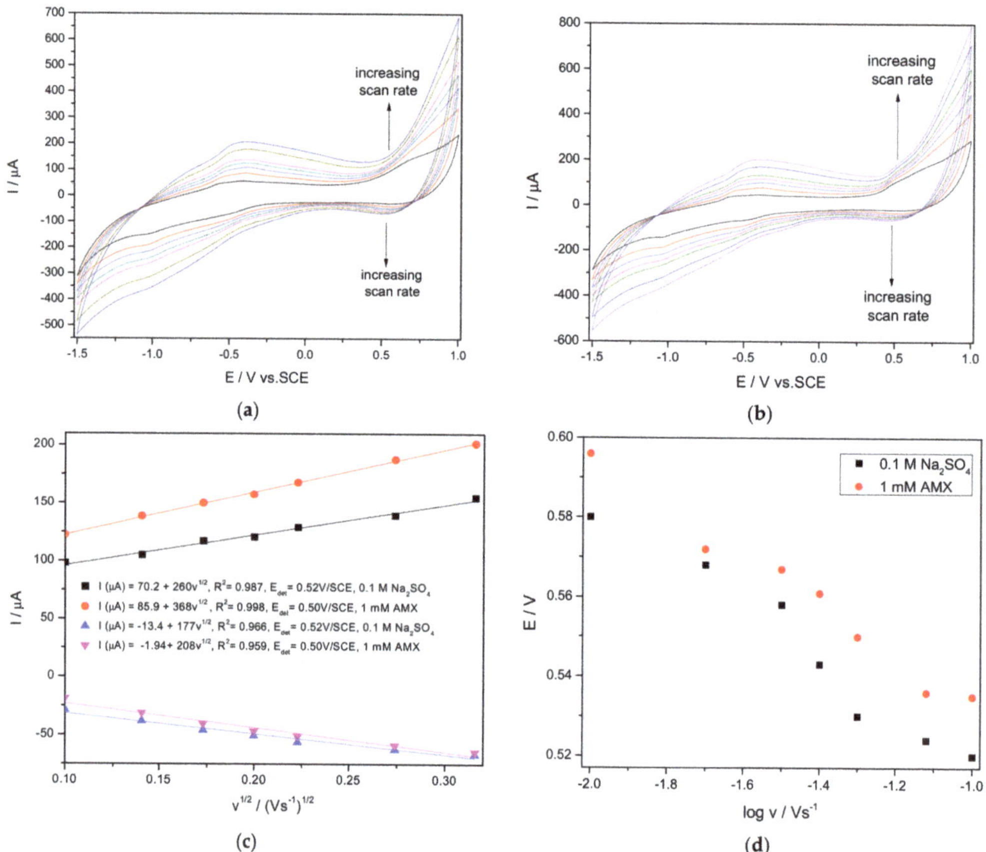

Figure 7. Cyclic voltammograms recorded at CuBi/CNF modified paste electrode in 0.1 M Na_2SO_4 supporting electrolyte (**a**) and 1 mM AMX (**b**) at various scan rates: (1) 10, (2) 20, (3) 30, (4) 40, (5) 50, (6) 75, and (7) 100 mVs^{-1}. Dependence of anodic peak current vs. square root of scan rate (**c**). Dependence of anodic and cathodic peak potentials vs. logarithm of the scan rate (**d**).

The electrochemical behaviour of CuBi/CNF in the absence and in the presence of AMX is similar to the shape. The anodic peak current recorded at +0.54 V/SCE increased linearly with the scan rate increasing. After the addition of 1 mM AMX, the anodic peak current significantly increased, while the corresponding cathodic peak decreased (see Figure 7c), which indicates that the AMX oxidation involves an electrochemical reaction coupled with a chemical reaction, which is an electrocatalytic reaction. No major changes in anodic peak potential with the scan rate increasing are noticed, which are slightly shifted to more negative values, indicating the redox system reversibility in both the absence and the presence of AMX (Figure 7d). It can be concluded that the active species responsible for the redox couple recorded at about +0.54 V/SCE plays an important role in electro-catalyzing the oxidation of AMX.

To enhance the electroanalytical performance for AMX detection, the electrochemical behaviour of the electrode in the presence of AMX and the operating characteristics of the voltametric and amperometric techniques are further considered.

3.3.3. Differential-Pulsed Voltammetry (DPV) and Square-Wave Voltammetry (SWV)

In comparison with CV, the electrochemical response provided by the differential-pulsed voltammetry technique is superior regarding the sensitivity, accuracy, and resolution for electrochemical sensing and electrochemical mechanism elucidation [36,37]. In AMX detection, DPV was tested under various operating conditions related to the step potential (SP), which ranged from 0.01 to 0.05 V while modulation amplitude (MA) ranged from 0.05 to 0.2 V. The shapes of DPVs are different when related to the operating conditions. The electroanalytical parameters are influenced by the electrochemical response stability. The best electroanalytical results were achieved for SP of 0.02 V and MA of 0.1 V corresponding to a 20-mVs^{-1} scan rate (Figure 8).

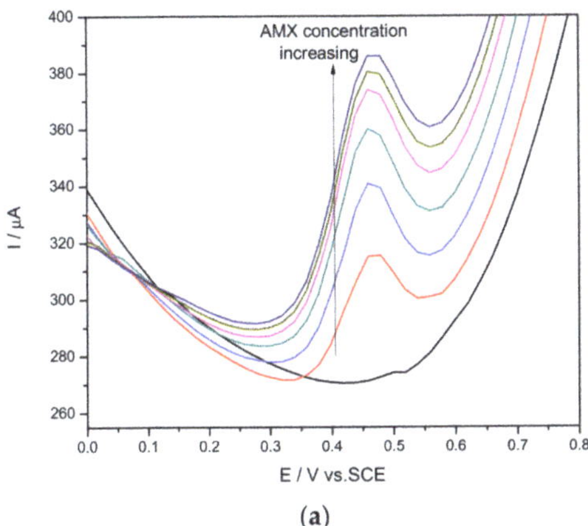

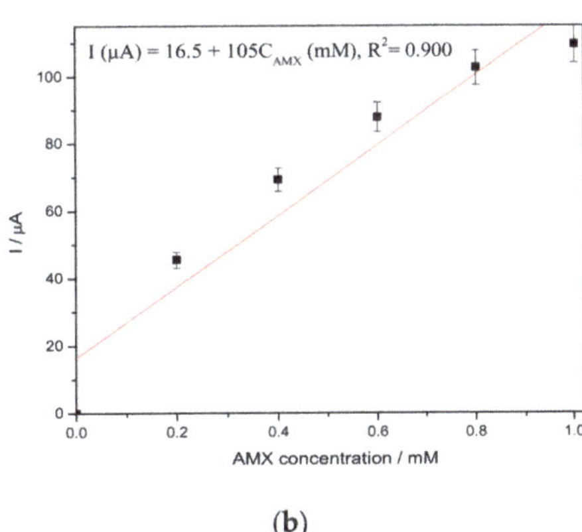

Figure 8. (**a**) Differential-pulse voltammograms recorded on the CuBi/CNF electrode in 0.1 M Na$_2$SO$_4$ supporting electrolyte in the presence of various AMX concentrations: 0.2 mM, 0.4 mM, 0.6 mM, 0.8 mM, 1.0 mM, and 1.2 mM. SP of 0.02 V and MA of 0.10 V. (**b**) Calibration plots for AMX detection in the concentration range of 0.2–1.2 mM recorded at the potential value of E = +0.500 V/SCE.

It can be noticed that the detection potential value is slightly shifted to a lower potential value in comparison with CV, which is one of the main characteristics of the differential voltammetry and is an advantage of this technique as related to the detection aspect. Under these operating conditions, the best sensitivity for AMX detection was reached at the potential value of +0.500 V/SCE (538 μA mM^{-1} cm^{-2}), which is much higher in comparison with one reached by CV (181 μA mM^{-1} cm^{-2}). Furthermore, the square-wave voltammetry (SWV) technique was tested for a comparison with DPV under the previously mentioned, optimized, operating DPV conditions. Taking into account the major advantage of SWV given by speed controlled by the frequency product (f) and step potential (SP), the frequency ranged from 10 to 50 Hz. The best results were achieved for the frequency of 20 Hz at the scan rate of 0.4 V·s^{-1} and the results are presented in Figure 9, which showed a stable and fast voltammetric response. All electroanalytical parameters reached for the voltametric techniques are shown comparatively in Table 2, and it can be noticed that DPV allowed reaching the lowest limit of detection of 1.5 10^{-7} M, while the best sensitivity (653 μA mM^{-1} cm^{-2}) was achieved by SWV.

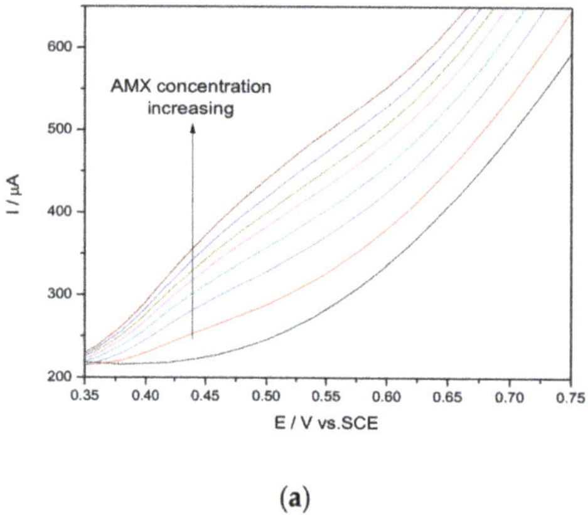

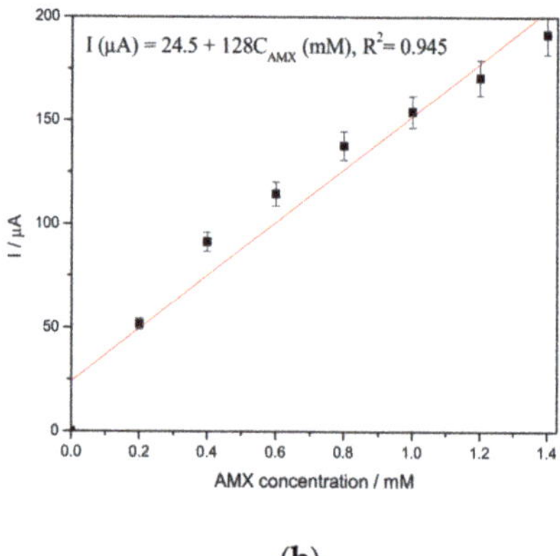

(a)   (b)

Figure 9. (**a**) Square-wave voltammograms recorded on CuBi/CNF electrode in 0.1 M Na$_2$SO$_4$ supporting electrolyte, in the presence of various AMX concentrations: 0.2 mM, 0.4 mM, 0.6 mM, 0.8 mM, 1.0 mM, 1.2 mM, and 1.4 mM. Step potential (SP) of 0.02 V, modulation amplitude (MA) of 0.1 V. f = 20 Hz. (**b**) Calibration plots for AMX detection in the concentration range 0.2–1.4 mM recorded at the potential value of E = +0.500 V/SCE.

Table 2. The electroanalytical parameters for AMX electrochemical detection using the CuBi/CNF electrode.

Technique	Working Parameters	Detection Potential (V/SCE)	Sensitivity (µA µM^{-1} cm^{-2})	LOD (µM)	LOQ (µM)	R^2
CV	v = 0.05 V s^{-1}	+0.550	181	0.965	3.22	0.938
		−1.00	78.5	1.31	4.80	0.920
DPV	SP = 0.02 V MA = 0.10 V V = 0.20 V s^{-1}	+0.500	538	0.150	0.520	0.946
SWV	SP = 0.02 V MA = 0.10 V f = 20 Hz; v = 0.20 V s^{-1}	+0.500	653	1.60	5.33	0.945
CA		+0.750	70.9	5.87	19.6	0.984
MPA	two potential levels, pulse time = 0.10 s	+0.750	503	2.43	8.12	0.983
		−1.00	183	4.74	15.8	0.937

3.3.4. Amperometry for AMX Detection on the CuBi/CNF Electrode

The amperometric techniques, considered as the simplest for practical applications, were tested in order to elaborate the enhanced amperometric detection protocol. This technique operates at one or more certain potential levels and the main disadvantage of chronoamperometry is the fast electrode fouling that means the loss of the amperometric signal. This aspect can be easily shown in sensitivity decreasing in comparison with CV results. Chronoamperograms recorded in the presence of the same AMX concentrations range (presented in Figure 10a) operated at one potential level of +0.750 V/SCE showed a sensitivity by about two times lower (Figure 10b) than the sensitivity obtained by CV due to possible electrode fouling.

In order to enhance the electroanalytical performance of amperometric detection of AMX, the pseudo multiple-pulsed amperometry (MPA) technique, operated at the two potential values corresponding to the reduction and the oxidation processes, was tested and the results of the amperograms and corresponding calibration plots are gathered in Figure 11a,b.

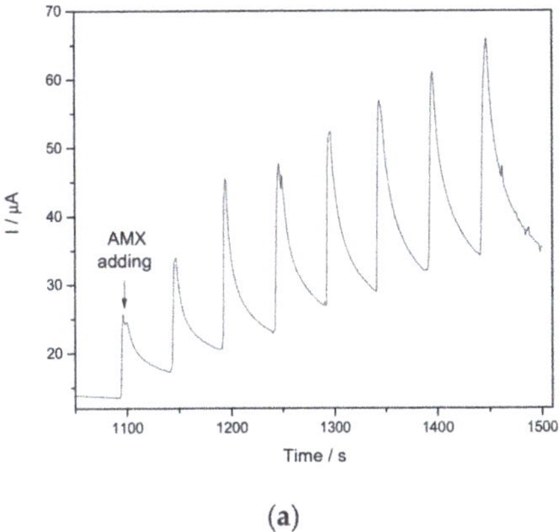
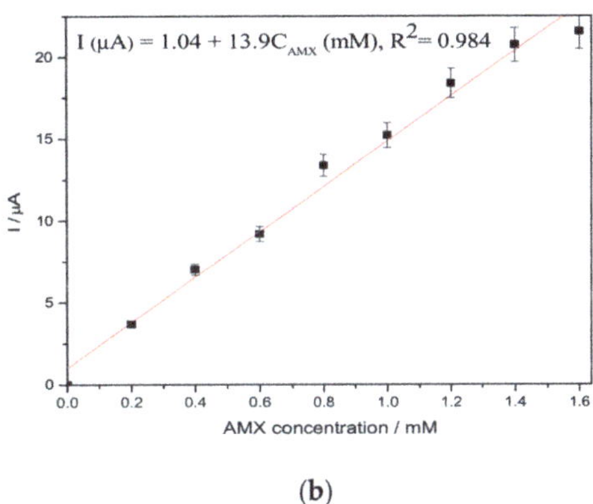

(**a**)

(**b**)

Figure 10. (**a**) Chronoamperograms recorded on CuBi/CNF electrode in 0.1 M Na$_2$SO$_4$ supporting electrolyte and in the presence of various AMX concentrations: 0.2 mM, 0.4 mM, 0.6 mM, 0.8 mM, 1.0 mM, 1.2 mM, and 1.4 mM at (**a**) an applied potential level of +0.750 V/SCE. (**b**) Calibration plots for AMX detection in the concentration range of 0.2 to 1.4 mM.

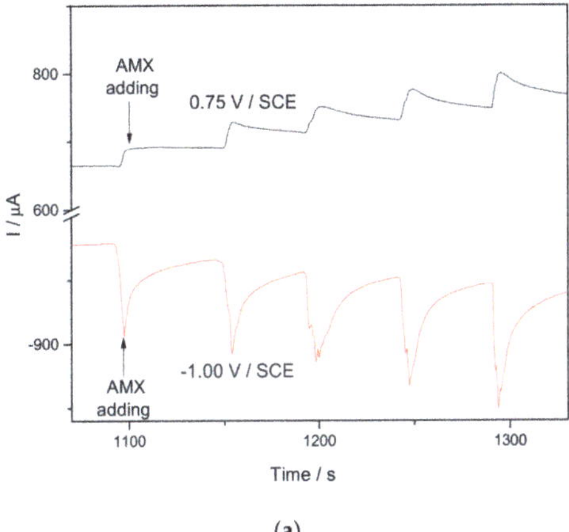
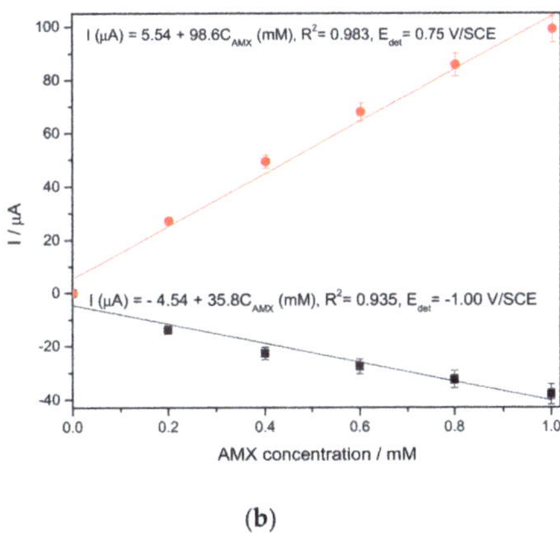

(**a**)

(**b**)

Figure 11. (**a**) Multiple-pulsed amperograms recorded at CuBi/CNF paste electrode in 0.1 M Na$_2$SO$_4$ supporting electrolyte in the presence of various AMX concentrations: 0.2 mM, 0.4 mM, 0.6 mM, 0.8 mM, and 1.0 mM, at two applied potential levels, i.e., +0.750 V/SCE and −1.00 V/SCE. (**b**)Calibration plots for AMX detection in the concentration range of 0.2–1.0 mM recorded at the potential +0.750 V/SCE and −1.00 V/SCE.

In general, it is well-known that the MPA technique exhibit the advantage of fouling avoiding for amperometric detection and is based on in-situ electrochemical activation of the electrode surface through fast and short amperometric pulses [38]. In this application, the MPA technique is adapted and named pseudo-MPA due to the electrode activation assured during the detection step without other supplementary pulses at the cathodic/anodic potential values, which are imposed for MPA. The anodic pulse level was selected at a potential value of +0.750 V/SCE, which is higher than the detection potential determined by CV and the cathodic one at the potential value of −1.00 V/SCE, in accordance with the CV results. The amperometric results are much better than voltammetric ones related to the sensitivity and it must be highlighted that the CuBi/CNF electrode potential for cathodic detection of AMX is a very promising aspect for simultaneous detection. The electroanalytical performance of detection achieved by the amperometry technique at the potential value of +0.750 V/SCE is comparable with one reached by the voltammetric techniques, which recommend this electrode and technique for further development of the amperometric-based protocol for AMX detection.

All detection results are presented in Table 2, and it can be noticed that CuBi/CNF is very promising for the AMX detection in aqueous solution.

The electroanalytical results related to the lowest limit of detection and the sensitivity obtained with the CuBi/CNF electrode are better than those reported by Essousi et al. [28], and showed the possibility of a practical application of this electrode in the detection of amoxicillin in the aqueous solution. The practical analytical application of the CuBi/CNF paste electrode using the DPV method was established by determining AMX in a water sample. Analyzing three parallel tap water samples spiked with 35 and 70 mg·L^{-1} AMX were selected for the recovery test. Good recovery and reproducibility of the results were found based on the minimum recovery values of 95% and the maximum relative standard deviation (RSD) values of 6% for both concentrations.

4. Conclusions

In this paper, $CuBi_2O_4$ powders were prepared by a new method based on the thermolysis of the oxalate coordination compound starting with 500 °C. XRD and SEM analyses indicated that phase-pure $CuBi_2O_4$ particles with an average particle size of 893.3 nm were obtained after calcination at 700 °C for 1 h.

The electrocatalytic effect of $CuBi_2O_4$ particles toward the amoxicillin detection was tested using a carbon nanofiber electrode as a substrate through the $CuBi_2O_4$/CNF paste electrode (CuBi/BDD) and in comparison with the CNF paste electrode. Cyclic voltammetric studies showed the superiority of CuBi/CNF paste through the redox system manifested within intrinsic $CuBi_2O_4$.

The voltammetric detection methods based on differential-pulsed voltammetry operated under 0.02 V as a step potential and 0.100 V as a modulation amplitude that allowed reaching the lowest limit of detection of 0.15 µM, while a fast and stable response characterized by the highest sensitivity of 653 µA µM cm^{-2} was achieved with square-wave voltammetry operated under a 0.02 Vas step potential and a 0.100 V as a modulation amplitude, frequency of 20 Hz, and a scan rate of 0.4 V s^{-1}. The amperometric detection method involving a pseudo multi-pulsed amperometry technique based on both anodic and cathodic potential levels led to very promising results for AMX detection related to sensitivity and selectivity. These results confirm the great potential of the CuBi/CNF paste electrode to be used for amoxicillin detection in the aqueous solution through the electrooxidation and electroreduction process, which should be further exploited for selective/simultaneous detection of amoxicillin in a multi-component matrix, considering a prior concentration stage that can be separate or included within the detection protocol by simple sorption onto the electrode surface.

Author Contributions: In this paper, R.D.(m.V.) and F.M. conceived and designed the experiments. C.P., S.N., A.P., and A.S. performed the experiments. R.D.(m.V.), F.M., S.N., A.S., and A.I. analyzed the data. R.D.(m.V.), F.M., and A.I. wrote the paper. All authors have read and agreed to the published version of the manuscript.

Funding: This work was supported partially by research grant GNaC2018-ARUT, no. 1350/01.02.2019, financed by Politehnica University of Timisoara and partially funded by a grant of the Romanian Ministery of Research and Innovation, CCDI-UEFISCDI, project number PN-III-P2-2.1-PED-2019-4492, contract number 441PED/2020 (3DSAPECYT), within PNCDI III.

Conflicts of Interest: The authors declare no conflict of interest.

References

1. Baciu, A.; Pop, A.; Manea, F.; Schoonman, J. Simultaneous arsenic (III) and lead (II) detection from aqueous solution by anodic stripping square-wave voltammetry. *Environ. Eng. Manag. J.* **2014**, *13*, 2317–2323.
2. Ardelean, M.; Manea, F.; Vaszilcsin, N.; Pode, R. Electrochemical detection of sulphide in water/ seawater using nanostructured carbon–epoxy composite electrodes. *Anal. Methods* **2014**, *6*, 4775–4782. [CrossRef]

3.   Marken, F.; Gerrard, M.L.; Mellor, I.M.; Mortimer, R.J.; Madden, C.E.; Fletcher, S.; Holt, K.; Foord, J.S.; Dahm, F. Voltammetry at Carbon Nanofiber Electrodes. *Electrochem. Commun.* **2001**, *3*, 177–180. [CrossRef]

4.   Motoc, S.; Manea, F.; Orha, C.; Pop, A. Enhanced Electrochemical Response of Diclofenac at a Fullerene–Carbon Nanofiber Paste Electrode. *Sensors* **2019**, *19*, 1332. [CrossRef]

5.   Manea, F.; Motoc, S.; Pop, A.; Remes, A.; Schoonman, J. Silver-functionalized carbon nanofiber composite electrodes for ibuprofen detection. *Nanoscale Res. Lett.* **2012**, *7*, 33. [CrossRef]

6.   Arai, T.; Yanagida, M.; Konishi, Y.; Iwasaki, Y.; Sugihara, H.; Sayama, K. Efficient complete oxidation of acetaldehyde into CO_2 over $CuBi_2O_4/WO_3$ composite photocatalyst under visible and UV light irradiation. *J. Phys. Chem. C* **2007**, *111*, 7574–7577. [CrossRef]

7.   Najafian, H.; Manteghi, F.; Beshkar, F. Fabrication of nanocomposite photocatalyst $CuBi_2O$ removal of acid brown 14 as water pollutant under visible light irradiation. *J. Hazard. Mater.* **2019**, *361*, 210–220. [CrossRef] [PubMed]

8.   Berglund, S.P.; Abdi, F.F.; Bogdanoff, P.; Chernseddine, A.; Friedrich, D.; van de Krol, R. Comprehensive Evaluation of $CuBi_2O_4$ as a Photocathode Material for Photoelectrochemical Water Splitting. *Chem. Mater.* **2016**, *28*, 4231–4242. [CrossRef]

9.   Kang, D.; Hill, J.C.; Park, Y.; Choi, K.S. Photoelectrochemical Properties and Photostabilities of High Surface Area $CuBi_2O_4$ and Ag-Doped $CuBi_2O_4$ Photocathodes. *Chem. Mater.* **2016**, *28*, 4331–4340. [CrossRef]

10.   Wu, C.H.; Onno, E.; Lin, C.L. CuO nanoparticles decorated nano-dendrite-structured $CuBi_2O_4$ for highly sensitive and selective electrochemical detection of glucose. *Electrochim. Acta* **2017**, *229*, 129–140. [CrossRef]

11.   Van Nguyen, T.H.; Cheng-Hsien, W.; Shao-Yu, L.; Chia-Yu, L. CoO_x nanoparticles modified $CuBi_2O_4$ submicron-sized square columns as a sensitive and selective sensing material for amperometric detection of glucose. *J. Taiwan Inst. Chem. Eng.* **2019**, *95*, 241–251. [CrossRef]

12.   Zhan, Y.; Lin, F.F.; Wei, T. Facile synthesis of Cu bismuthate nanosheets and sensitive electrochemical detection of tartaric acid. *J. Alloys. Compd.* **2017**, *723*, 1062–1069. [CrossRef]

13.   Guo, X.Y.; Mao, Y.J.; Yu, C.H.; Qiu, F.L.; Pei, L.Z.; Ling, X.Z.; Zhang, M.C.; Fan, C.G. Polythiopene/copper bismuthate nanosheet nanocomposites modified glassy carbon electrode for electrochemical detection of benzoic acid. *Int. J. Electrochem. Sci.* **2020**, *15*, 10463–10475. [CrossRef]

14.   Zhang, Y.C.; Yang, H.; Wang, W.P.; Zhang, H.M.; Li, R.S.; Wang, X.X.; Yu, R.C. A promising supercapacitor electrode material of $CuBi_2O_4$ hierarchical microspheres synthesized via a coprecipitation route. *J. Alloys Compd.* **2016**, *684*, 707–713. [CrossRef]

15.   Muthukrishnaraj, A.; Vadivel, S.; Made Joni, I.; Balasubramanian, N. Development of reduced graphene oxide/$CuBi_2O_4$ hybrid for enhanced photocatalytic behavior under visible light irradiation. *Ceram. Int.* **2015**, *41*, 6164–6168. [CrossRef]

16.   Xie, Y.; Zhang, Y.; Yang, G.; Liu, C.; Wang, J. Hydrothermal synthesis of $CuBi_2O_4$ nanosheets and their photocatalytic behavior under visible light irradiation. *Mat. Lett.* **2013**, *107*, 291–294. [CrossRef]

17.   Chen, X.; Dai, Y.; Guo, J. Hydrothermal synthesis of well-distributed spherical $CuBi_2O_4$ with enhanced photocatalytic activity under visible light irradiation. *Mat. Lett.* **2015**, *161*, 251–254. [CrossRef]

18.   Chen, X.Y.; Ma, C.; Li, X.X.; Chen, P.; Fang, J.G. Hierarchical Bi_2CuO_4 microspheres: Hydrothermal synthesis and catalytic performance in wet oxidation of methylene blue. *Catal. Commun.* **2009**, *10*, 1020–1024. [CrossRef]

19.   Hossain, M.K.; Samu, G.F.; Gandhan, K.S.; Santhanagopalan, J.; Ping Liu, C.; Rajeshwar, K. Solution combustion synthesis, characterization, and photocatalytic activity of $CuBi_2O_4$ and its nanocomposites with CuO and α-Bi_2O_3. *J. Phys. Chem. C* **2017**, *121*, 8252–8261. [CrossRef]

20.   Anandan, S.; Lee, G.; Yang, C.; Ashokkumar, M.; Wu, J.J. Sonochemical synthesis of Bi_2CuO_4 nanoparticles for catalytic degradation of nonylphenol ethoxylate. *Chem. Eng. J.* **2012**, *183*, 46–52. [CrossRef]

21.   Fagerquist, C.K.; Lightfield, A.R.; Lehotay, S.J. Confirmatory and quantitative analysis of beta-lactam antibiotics in bovine kidney tissue by dispersive solid-phase extraction and liquid chromatography tandem mass spectrometry. *J. Anal. Chem.* **2005**, *77*, 1473–1482. [CrossRef]

22.   Lima, D.R.; Lima, E.C.; Umpierres, C.S.; Thue, P.S.; El-Chaghaby, G.A.; da Silva, R.S.; Pavan, F.A.; Dias, S.L.; Biron, C. Removal of amoxicillin from simulated hospital effluents by adsorption using activated carbons prepared from capsules of cashew of Para. *J. Environ. Sci. Pollut. Res.* **2019**, *26*, 16396–16408. [CrossRef]

23.   European Commission. Commission Implementing Decision (EU) 2018/840 of 5 June 2018 establishing a watch list of substances for Union-wide monitoring in the field of water policy pursuant to Directive 2008/105/EC of the European Parliament and of the Council and repealing Commission Implementing Decision (EU) 2015/495. *Off. J. Eur. Union* **2018**, *L 141*, 9–12.

24.   Matar, K.M. Simple and rapid LC method for the determination of amoxicillin in plasma. *J. Chromatogr.* **2006**, *64*, 255–360. [CrossRef]

25.   Pajchel, G.; Pawlowski, K.; Tyski, S. CE versus LC for simultaneous determination of amoxicillin/clavulanic acid and ampicillin/sulbactam in pharmaceutical formulations for injections. *J. Pharm. Biom. Anal.* **2002**, *29*, 75–81. [CrossRef]

26.   Hrioua, A.; Loudiki, A.; Farahi, A.; Bakasse, M.; Lahrich, S.; Saqrane, S.; El Mhammedi, M.A. Recent advances in electrochemical sensors for amoxicillin detection in biological and environmental samples. *Bioelectrochemistry* **2021**, *137*, 107687. [CrossRef] [PubMed]

27.   Rezaei, B.; Damiri, S. Electrochemistry and Adsorptive Stripping Voltammetric Determination of Amoxicillin on a Multiwalled Carbon Nanotubes Modified Glassy Carbon Electrode. *Electroanalysis* **2009**, *21*, 1577–1586. [CrossRef]

28. Essousi, H.; Barhoumi, H.; Karastogianni, S.; Girousi, S.T. An Electrochemical Sensor Based on Reduced Graphene Oxide, Gold Nanoparticles and Molecular Imprinted Overoxidized Polypyrrole for Amoxicillin Determination. *Electroanalysis* **2020**, *32*, 1546–1558. [CrossRef]

29. Karuwan, T.; Mantim, P.; Chaisuwan, P.; Wilairat, Y.; Einaga, O.; Chailapakul, L.; Suntornsuk, L. Pulsed Amperometry for Anti-fouling of Boron-doped Diamond in Electroanalysis of β-Agonists: Application to Flow Injection for Pharmaceutical Analysis. *Sensors* **2006**, *6*, 1837–1850. [CrossRef]

30. Dumitru, R.; Manea, F.; Păcurariu, C.; Lupa, L.; Pop, A.; Cioablă, A.; Surdu, A.; Ianculescu, A. Synthesis, Characterization of Nanosized $ZnCr_2O_4$ and Its Photocatalytic Performance in the Degradation of Humic Acid from Drinking Water. *Catalysts* **2018**, *8*, 210. [CrossRef]

31. Fujita, J.; Nakamoto, K.; Kobayshi, M. Infrared Spectra of Metallic Complexes. II. The Absorption Bands of Coördinated Water in Aquo Complexes. *J. Am. Chem. Soc.* **1956**, *78*, 3963–3965. [CrossRef]

32. Dumitru, R.; Papa, F.; Balint, I.; Culita, D.; Munteanu, C.; Stanica, N.; Ianculescu, A.; Diamandescu, L.; Carp, O. Mesoporous cobalt ferrite: A rival of platinum catalyst in methane combustion reaction. *Appl. Catal. A Gen.* **2013**, *467*, 178–186. [CrossRef]

33. Nakamoto, K. *Infrared and Raman Spectra of Inorganic and Coordination Compounds*; Wiley: New York, NY, USA, 1986; ISBN 0471010669.

34. Dumitru, R.; Ianculescu, A.; Păcurariu, C.; Lupa, L.; Pop, A.; Vasile, B.; Surdu, A.; Manea, F. $BiFeO_3$-synthesis, characterization and its photocatalytic activity towards doxorubicin degradation from water. *Ceram. Intern.* **2019**, *45*, 2789–2802. [CrossRef]

35. Konopka, S.J.; McDuffie, B. Diffusion coefficients of ferri- and ferrocyanide ions in aqueous media, using twin-electrode thin-layer electrochemistry. *Anal. Chem.* **1970**, *42*, 1741–1746. [CrossRef]

36. Vilas-Boas, A.; Valderrama, P.; Fontes, N.; Geraldo, D.; Bento, F. Evaluation of total polyphenol content of wines by means of voltammetric techniques: Cyclic voltammetry vs differential pulse voltammetry. *Food Chem.* **2019**, *276*, 719–725. [CrossRef]

37. Farahani, K.Z.; Benvidi, A.; Rezaeinasab, M.; Abbasi, S.; Abdollahi-Alibeik, M.; Rezaeipoor-Anari, A.; Zarchi, M.A.K.; Abadi, S.S.A.D.M. Potentiality of PARAFAC approaches for simultaneous determination of N-acetylcysteine and acetaminophen based on the second-order data obtained from differential pulse voltammetry. *Talanta* **2019**, *192*, 439–447. [CrossRef] [PubMed]

38. Pop, A.; Lung, S.; Orha, C.; Manea, F. Silver/graphene-modified boron doped diamond electrode for selective detection of carbaryl and paraquat from water. *Int. J. Electrochem. Sci.* **2018**, *13*, 2651–2660. [CrossRef]

MDPI
St. Alban-Anlage 66
4052 Basel
Switzerland
Tel. +41 61 683 77 34
Fax +41 61 302 89 18
www.mdpi.com

Nanomaterials Editorial Office
E-mail: nanomaterials@mdpi.com
www.mdpi.com/journal/nanomaterials